BIG SNOWY MTS.

LITTLE BELT MTS.

JUDITH GAP

HARLOWTON

TWO DOT

MUSSELSHELL RIVER

CRAZY MOUNTAINS

COFFIN BUTTE

BIG ELK CREEK

AMERICAN FORK

FISH CREEK

PORCUPINE
BUTTE

CAYUSE HILLS

BIG COULEE

BUTTE RANCH

SWEET GRASS CREEK

DRY CREEK

OTTER CREEK

MELVILLE

"NO BUSH"

BLACK BUTTE

HOME RANCH

VAN CLEVE'S
DUDE RANCH

WHEELER CREEK

CRAZY MOUNTAINS

SOUTH FORK

10 MILE CREEK

BIG TIMBER CREEK

SOUR DOUGH

RASPBERRY BUTTE

YELLOWSTONE RIVER

BIG TIMBER

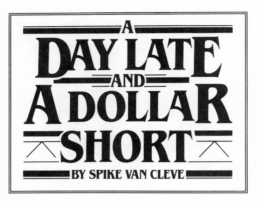

A DAY LATE AND A DOLLAR SHORT

BY SPIKE VAN CLEVE

Nothing and no one came before Barbara

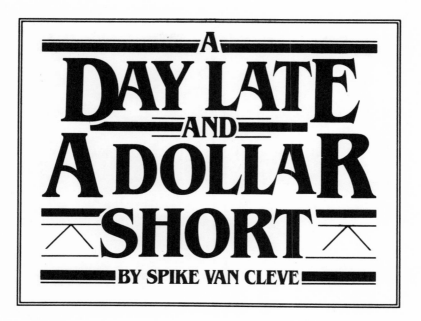

A DAY LATE AND A DOLLAR SHORT

BY SPIKE VAN CLEVE

Introduction by
A. B. GUTHRIE, JR.

THE LOWELL PRESS/KANSAS CITY

Also by Spike Van Cleve

40 YEARS' GATHERIN'S

Library of Congress Cataloging in Publication Data
Van Cleve, Spike.
A day late and a dollar short.
1. Van Cleve, Spike.
2. Sweet Grass County (Mont.)—Social life and customs.
3. Ranch life—Montana—Sweet Grass County.
4. Ranchers—Montana—Sweet Grass County—Biography.
5. Cowboys—Montana—Sweet Grass County—Biography.
6. Sweet Grass County (Mont.)—Biography.
I. Title.
F737.S9V37   978.6'64   81-83817
ISBN 0-913504-65-3   AACR2

FIRST EDITION
Fifth Printing, 1991

Printed in the United States of America by The Lowell Press, Inc.
Kansas City, Missouri.

*To my family
and
to my friends*

Spike on Peep O'Day at Upper Twin Lake (summer cow range) in the Crazies

CONTENTS

Spike with his beloved horse Sam Hill on the south slope of the Big Hill. The photograph was used for his article in the August/September 1979 issue of CLASSIC Magazine.

INTRODUCTION

Dear Spike:

I have read the galley proofs of your new book. Your words are like a warm and engaging letter, addressed to me alone. I am sure a great many Montanans and westerners will feel the same way. Since you have written to me, I shall answer by letter, wishing like hell to bridge the chasm that separates us now.

It was maybe two weeks ago that you telephoned, asking if I would write the introduction to your new book. There was a quick answer to that. I said, "Sure, Spike, send the proofs along."

Asked about yourself, you said you were fine except for a blood clot that had climbed from your lower leg to your upper. You told me you weren't answering well to blood-thinning drugs. I thought then of hearts and embolisms, though I didn't say so, and asked you to take care.

Within three or four days I had the sorry news that you had died. To men like us there is only one reply, a barnyard reply, to that kind of news, and I said it aloud and to myself.

We met only once, as you will recall. That was at a shindig for writers at Billings. I suspect, as the hardboots of Kentucky would say, that you felt like a plater in a Thoroughbred handicap. No matter. You weren't fazed. In the presence of better-established writers you were your unassuming but independent self. You knew, by God, who

*Spike on his horse
Streak, about 1932*

Spike with one of his at-halter colts in Bozeman, Montana

you were and where you stood. There are not enough of that breed around.

So we met only once, but I knew right away, as Kipling would have said, that we were of one blood. And not alone or even because you liked what I had written, though you said, "By God, Bud, when your western characters speak, they talk like the men I know and work with." There was more to our encounter than that—a sort of recognized relationship, the warmth of common experience, of shared attitudes.

By comparison with you, I am only a tenderfoot, but, Spike, I have ridden many a horseback mile in sub-zero weather, burdened by the heavy clothing required. I have known many winters and winds. Remember telling me that, when a dude asked you if you didn't hate the wind, you answered that it swept the ranges clear of snow and let the cows graze? You were right on target. It's the wind that helps make Montana a great cattle country.

You use a happy description in speaking of Melville humor. You call it "slaunchwise." That kind of humor is common to Montana and perhaps the interior northwest. Here's a Teton-county example that you will like. Once a friend of mine was taken with drink, as you would say. Next morning his wife remarked, "You came home with quite a load on last night, didn't you?" To which he replied, "Yep, should have made two trips."

The old west bred storytellers. Books were few then, and literacy hardly universal. Those who could read tired of the labels on tomato cans. In the additional absence of radio and TV, the man who could embellish a tale drew an audience. Two of the best of these entertainers were Con Price and Charlie Russell. I add you to the list. You belong in that company. And I hope you are not the last of the breed.

The eastern-seaboard veneer didn't stick to you worth a damn during your studies at Harvard. But, outside Melville, something somewhere did—an attachment to reading, a liking for words, an ability to line them up and make them work. Whoa, now I quote you:

> *Or perhaps there'd be a vicious drift of air out of the northeast (which I'd have to quarter into on my way home) with the sky and ground completely indistinguishable. Scurrying little trickles and snakes of snow would appear in the white ahead, whisper past and disappear into the white behind us. Only the Crazies and Porcupine showed their tops above the frost haze—unreal, bitter and icy blue.*

That's writing. I not only get the picture: I'm in it.

Your whole book—be embarrassed then—adds to our lore and our entertainment and for longer than you can believe will be consulted by students who want to know how we lived, how we spoke and what our attitudes were.

I hope God gives you a horse.

Your friend,

Bud

A. B. GUTHRIE, JR.

Choteau, Montana
April 22, 1982

Spike at ease at the saddle shed and corrals at the dude ranch

Spike in camp on a camp trip with Barbie on the Breaks of the Missouri (Charlie Russell Wildlife Refuge)

Typical Spike

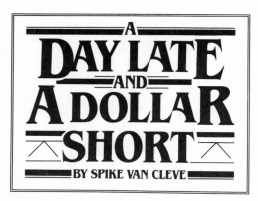

A DAY LATE AND A DOLLAR SHORT

BY SPIKE VAN CLEVE

=1=
THE PURE QUILL

I SMILE FONDLY whenever I think about Dad. Especially when he got old and crippled, for he was such a hell of a man! At best, he never was in the lead when it came to an even temper, and he got worse towards his last years. I don't blame him, for he only had one eye, couldn't hear very well and was so stove up in the knees from work and rank horses that he couldn't get around without the help of a cane—two, if he happened to remember them both.

What frustrated him the worst, I think, was that he no longer could do what he called "a day's work," which for an average man would be anyhow a two-day job. If it hadn't been for his Jeep and his two dogs, I don't know what he'd have done, for he'd take those three and work cattle like you wouldn't believe—him in a gate up on the range, one dog bringing the stock to him and the other holding the cut. Four men couldn't have done any better, and, one-eyed or not, he sure could read a brand. Besides, he knew every damn cow he owned and every quirk she had. Oh, he got hooked a few times by rank old girls, and once in a while he'd come into the ranch in reverse, having run into a fence somewhere and gotten so much wire wrapped around the drive shaft it would only work in one direction. But he was happy and busy right up to the spring of '69.

That spring he was in bad shape. Couldn't get out of the house, let alone drive his Jeep. So one bright warm day, Barbara, who is the kindest, most sensitive person I know,

1

suggested, "Spike, why don't you go down in the pickup and see if Dad would like to take a ride with you? He'd love to see the calves and the colts. The grass is showing green, the creeks are singing, you said you'd seen a curlew the other day, and I know he'd enjoy it. It might be the last time, honey."

It was. He never got out of the house again, but the two of us had a great time together; and when he thanked me at his door, there were tears in his eyes. "It was Barbara's idea, Pop," I told him, "and I'm sure glad she thought of it."

"Hell," he grinned, "I knew that. Thank her for me and give her my love." About a week later he was dead.

I always figured Dad could ride a horse to death that I wouldn't have gone into the corral with. Except maybe for his last bronc ride. But it sure was so, up until that last ride, and the tough old rascal was seventy-seven years old when he made it.

It came about because the horse he'd used for years, a big, ouchy chestnut gelding that I wouldn't have ridden if he'd given him to me, Colonel, had finally stiffened up, and we'd had to put him away. The next spring Dad spotted a stout chestnut stud colt when he hit the ground and remarked, "There's my new horse. I'll call him Colonel when I get him broken."

Didn't matter a bit that Pop was over seventy, stove up something terrible, and that the colt was out of a raspy Thoroughbred mare we'd given up on and put in the brood mare bunch. Hell, no, not to Dad! So when the colt was four, he started him, and be damned if he didn't get him broken.

Still, I had a feeling and didn't trust the horse. So one day, as Dad caught him when we were branding a string of big calves, I told him, "Dammit, you can't rope on that colt. He's plumb green, and owly besides. Look, if you figure he needs work, why not let me ride him?"

"You're too stove up," he answered—I'd had a wreck with an old pony a few days before—and he so crippled himself he couldn't get around afoot!

"All right, then. I've got a couple of boys here that would be glad to ride him. How about one of them?"

"I ride my own horses, by God," he returned grimly, so I gave up.

Things rocked along in good shape, and Pop heeled calves with the best of us. It was a big corral—an arena, really—and as I was dragging a calf to the fire from the yon end, I heard a yell behind me and looked back in time to see the big colt break in two.

He sure turned the crank, but Dad was fitting a bronc ride—though as I spurred towards them, I could see he was getting looser each jump. I counted eight of them when, just before I got there, Dad came down, and, Lord, but he came down hard! Right flat on his back, and every inch from his head to his spurs hit at the same time—and bounced. He was trying to get his wind back when I stepped down and lifted him against my chest. "You all right, Dad?" I asked anxiously. "He hurt you?"

"You damn right he hurt me!" He shook his head. "I guess that does it. Should have quit sooner. Just bullheaded."

"Hell, Pop, that colt can buck," I assured him as I helped him up.

"Yep, noticed that myself," he grunted. It was the last time Dad was ever ahorseback. Two years later he was dead.

Twenty years earlier he'd have ridden the horse and whipped him over and under with his catch rope till he quit, for he sure was a rough horse hand. I came in with him from irrigating one day when he saw the horse breaker he'd hired—we still were breaking quite a string of colts every spring—roughing the living hell out of a big five-year-old. Doing it afoot, not on him.

"Hey," Dad called, "don't wool that horse around like that. It'll make him sour. Just ride him."

"Maybe you'd like to show me how you want it done," was the answer from the corral.

Pop was crowding sixty along about then but said, "I'll sure give it a try."

He leaned his shovel against the corral, climbed over it, headed for the horse, pulling the tops of his gum boots up around his legs as he went. "Give him here," he said, and, "Spike, toss me those rawhide hobbles."

Aboard he got, and the ball commenced. He got a little loose a time or two but never stopped whipping with the hobbles, and pretty quick the colt quit bucking and went to running, all the fight gone. After a few circles around the pen, Dad pulled up, stepped off and turned to the open-mouthed horse breaker. "That's what I want you to do when they need it. From on top of them," handed him the reins and climbed out of the pen, irrigating boots and all.

I remember one time, maybe twenty-five years ago, when Dad, Tor and I were working a string of cattle up under the mountains: two-year-old black heifers and, naturally, as giddy as a bunch of high school girls. On top of that, the spot Dad picked to work them was the bottom of a steep draw at the head of Billy Creek in amongst diamond willows, boggy spring seeps, with a lot of rocks the size of a man's head, or bigger, hidden in the long grass. A half-mile below we'd have been on a good, open flat, but we'd found these goosey boogers brushed up in the willows on water, and, as far as Dad was concerned, we were going to make our cut right, by God, there. We always claimed, but not too loudly, that there were three ways to do anything: the easy way, the hard way, and Dad's way— which made the hard way look easy! So we had unadulterated hell.

Pop was riding a bay gelding we called Sid. A good cowhorse, but he had a little age on him; and the way we had to ride, plus the conditions, were pretty rough on the old fellow. Besides, Dad warmed up and was really getting after him, spurs and reins both, whenever they had to make a run. Which was right frequent.

This went on for a while, and we had the job about done when all of a sudden, as Dad went to make a bust at a bunch-quitter, the bay dropped his head, let out a bellow that'd have put the run on a grizzly and broke in two.

Damn, but he did blow! Through, over, and into brush, bog, boulders and cattle he went, Dad dodging things, cussing and spurring when he got the chance. Which wasn't too often. The heifers exploded, a couple hit the cut and away they went, too. Tor and I did our damndest but we spilled the whole outfit, so we turned back to the uproar of breaking brush and loud bucking where Sid was still hard at it. Pop wasn't doing much spurring about then and, finally, damned if the old bay didn't bed him down. Hell, he must have bucked for anyhow three minutes or more before he got the job done, but he did it fair and square. Then he stopped and stood, sweating and shaking, but with his ears up. Everything about him seemed to say, "There, damn you!"

Dad got up, walked over, got the reins, and the two eyed one another. And it was about an even break who was the most played out. Then, gently, he patted the wet forehead. "Sorry," he said, "I was rougher on you than I had any business being. I had that coming."

Turning to us he grinned. "The old devil got mad and flat made up his mind to buck me off. I got after him awful bad when he was already doing his very best. Don't blame him, either. Matter of fact, damn if I don't respect him for it."

He started to loosen the cinches. "You two go after those cattle. Throw them into the fence corner down the creek, and when this old fellow gets his wind back, we'll be along and finish the job."

It didn't take long to get the cutting done when we got

the stock bunched, for they'd had a nice run to sort of take the edge off, and Dad and Sid worked together like nothing had ever happened, like the cowman and cowhorse they were. I noticed, too, that the old bay got an extra bait of grain when we unsaddled at the barn that evening.

Cowpunchers are a breed that thrives on excitement, and Dad was damn sure typical. His long suit was for horses that would provide it, and he had them. I know several instances where, if it were the only way he could get a sitting at some salty pony with a reputation, he'd buy him. Or trade for him, or something.

When I was in my teens, he had a horse of that caliber. A big, active, well put together grey gelding, obviously well-bred. We called him Rattler, but the name didn't begin to fit what a dirty, bucking, stampeding, kicking, striking, biting, treacherous snake he was. I think even Pop got sort of tired of it finally, for one day up on the Dry Creek flats, he lost his temper and did what, in my considered opinion, was about the wildest thing I ever saw a man do ahorseback.

We were under the rimrock on the south side of Battleship Butte, Dad riding Rattler. Suddenly, and for no reason at all, the grey bawled and built to it down the steep slope, through, over and around the chunks of rim that had fallen from above. High, wild and handsome! Dad never turned a hair. Just spurred from hell to breakfast. As they hit the flat, still in one piece and upright, he leaned ahead, slipped the one-ear bridle off and went to whipping with it. Both sides, back where a colt sucks its mammy. "You grey sonofabitch," he gritted, "you want to buck, buck," and I could hear the thud of the bit and the lash of the reins under the flanks.

Godamighty! To our left the flats dropped off into Otter Creek. Ahead of us they were cut by a steep, deep, partially rimrocked coulee that would make a goat look at his hole card, and the horse was bucking straight for it, plumb free and blind mad. I tried to ride in and turn them enough to miss the coulee, but my pony sure wanted no part of what was going on. I accomplished about as much as the one-legged guy at the butt-kicking contest, and, damn, but the back of my neck got cold!

Then Dad took care of things. Without missing a lick with the bridle, he switched from the grey's flanks to the near side of his head, and I could hear the bit ring like a bell against his jaws. It must have gotten the bronc's attention after a few busts, maybe because he was trying to get away from it or because it softened him up enough to take note of what was ahead, but anyhow they swung enough to the offside that they cleared the edge of the drop-off, and my hair let my hat back down on my head.

Even so, things weren't all that shiny yet. Though once past the coulee head the flats above Otter Creek veered off to the south some, for a half-mile or so they were topped by a rimrock anywhere from twenty to forty feet high. Besides, time he got by the coulee, Rattler had evidently decided pitching wasn't his plumb long suit and maybe a good run would be the ticket, so up the flat he stampeded. "You want to run," Dad told him, "then run, damn you," and switched back to the flanks again.

He had to bend the pony away from the rim a time or two, like he had done at the coulee, and when I saw the horse come around, I let my weight down a little. Lord God, if they'd hit that rim at the speed they were traveling, they'd have resembled ski jumpers. Until they lit, that is.

We went anyhow a mile up the flat as hard as the grey could run, Dad thumping him all the way. Finally the runaway stopped, plumb played out. Pop swung down, put the bridle back on and asked, "How'd you like it, you grey bastard?" Then he looked up at me, saying "I thought I might have to quit him for a while there. Then I'd have had to start all over again. This way maybe he's learned something."

I guess he had, for he never bucked again, even after he
healed up. Turned over a new leaf and got to be a good
horse, but it struck me, and still does, as a pretty juicy
mode of equine education. And it took a damn juicy man to
do the educating.

Pop had a trick with nasty stock, too, that I never saw
anywhere else. Particularly if a bronc liked to blow up or
spook by and kick, which can be damn rough on the
kickee's legs or stomach—as I know from bitter experience.
He'd take a fairly short rein and the saddle horn in his left
hand, step into the near stirrup and stand, right hand
holding the off side of the cantle. Then, if the old pony tried
anything gaudy, Dad would kick him in the belly with his
free leg till he quit. Sounded like somebody beating a bass
drum, but it usually didn't take long to get the wrinkles
ironed out of a bad actor, though they might take a real
dim view of it at first. They never forgot, either, and a horse
that'd had the treatment stood for a man to get on, from
then on out, like he'd grown roots. Pop claimed, "Hell,
there's no way they can buck you off. You've got two good
handholds. And if one turns over, you are out where you
just step down. No way they can catch you. You've got the
percent on them, no matter what."

Maybe so, but it always seemed to me that he was a little
sanguine. In any event, I never got the real hang of it and
found that it *was*, by God, possible to buck off, two
handholds and all.

Back when if over a half-dozen men bunched up in the
Melville country, there'd be a bucking contest, Dad had
quite a reputation as a hard man to beat. By the time
rodeo, as such, got up here, he was past his prime as a
saddle bronc rider but took to calf roping and bulldogging
like a wolf pup takes to red meat. As a matter of fact, he
entered the latter event at Madison Square Garden in '31. I
don't believe he won any money, but Dave Campbell, who
qualified as quite a hand himself, used to shake his head

over something Dad did at the Garden 'dogging. Seems his hazer crowded the steer under Pop's horse's neck to where he had to get off on the yon side of the steer. He got a horn, vaulted over the neck to the near side and threw the animal. In decent time, too. Dave never forgot that deal, nor did Oral Zumwalt. Zummy told me years later, "Spike, your Dad was about as salty an all-around, any time, anywhere, anything top cowboy as I've ever run into. Not a rodeo hand, but damn sure a cowboy. From who laid the chunk!"

Pop grew up roping big cattle, for steer ropings were the rule up here, and all his life he threw a soft, fairground loop. Even so, he took part of a calf roping day money at the Tucson La Fiesta de Los Vaqueros in '38. Back when *every* entrant got his stock *every* day, it was *the* roping rodeo, and there were at least seventy-five of the finest ropers in the world entered in the event.

They used to open the Tucson show with a bull riding scramble—a bull and rider out of every chute, all at the same time. The top bull riders steered clear of a mess like that—unless they happened to be damn badly broke—so the entrants were mainly amateurs. Happened this day everybody bucked off but one young fellow whose mount, when he quit firing, joined the loose cattle and, damn, but the boy looked lonely in the crowd, abullback all by himself. The clown wanted no part of that many owly Brahmas all at once, so the pickup men finally started to put the whole kit and caboodle into the catch pen. The cattle were real ouchy, and they were getting nowhere in a hurry when Dad, who'd been sitting on his rope horse watching, said, "If that boy can't get off out in the arena, how the hell do they figure he'll be able to when everything's crowded into a corral? He's liable to get hurt." And he drove into that boiling, hooking crowd, picked up the boy and brought him out safely. As he did, some young, hotshot roper remarked loudly, "That damned old fool is likely to get killed taking chances like that."

Earl Thode eyed him, "You watch that 'old fool' close, gunsel, and you just might learn something, though I doubt it like hell. *There's* a cowboy; not just somebody dressed like one!"

I honestly don't think Dad knew the meaning of fear. For himself, that is. He could get spooked for someone else, sure, but I never saw or heard of an instance when he gave a hoot in hell about what might happen to himself. Maybe that's why he could get away with some of the things he did. Like one time, when he and one of the boys were over working on a "company" ditch out of the Sweet Grass. I wasn't there, but the other guy told me the story, quotes and all. Said he'd never forget it.

The two were shoveling out a bank cave-in when a car drove up and a man, a comparative newcomer to the Melville country who'd bought a ranch with rights on the ditch, stepped out. He'd done some talking concerning what he was about to do to Paul Van Cleve, though why, I be damned if I know. Nobody paid him much mind, for talk's cheap. To boot, it had been tried a number of times over the years with damn little success. This man, though, was packing a .45 automatic on full cock, and before anyone savvied what was up, he poked the thing in Dad's belly and announced, "I'm gonna kill you." Just like that!

Pop straightened, sized up the deal and answered levelly, "Hop to it. Then my troubles'll be over, and yours'll commence."

They stood there a minute with the man's finger white on the trigger. Then Dad said coldly, "Well, get to it. Either touch that thing off or get back in your car and get the hell out of here, because if you don't do one or the other damn quick, I'm going to bend this shovel over the top of your head."

The man backed off, climbed in his car and left! The fellow with Pop told me it scared the hell out of *him*, and the hole in the end of that gun resembled the mouth of a nail keg. I asked Dad a while later if *he* hadn't been spooked, and I'll always remember his answer. "I don't know, really. I guess I should have been, but there wasn't a damn thing I could do about it one way or the other, and it made me mad. Everything I said was sure true, too."

Dad was just as unafraid of death when he got old and in bad shape as he'd been when he was in his prime. Damndest man! I dropped into the home ranch, not too long before he died, to say howdy. We were talking about the colt crop, the youngsters that were being broken and how the saddle horses had wintered when, out of a clear sky, he remarked casually, "You know, I bet I have a big funeral."

A little taken aback, I agreed and added something about how many friends he had. "I know," was his answer, "but there'll be a hell of a lot of people there, too, just to make damn sure I'm really dead."

The hell of it is that I never fully appreciated Pop when he was alive. Only after he was gone, I began to really know him as he was. We didn't see eye to eye on a lot of things when he was alive, especially after I grew up and was running my own outfit. But now something happens, and I think, "Dad'll get a boot out of this," or "Pop would have done such and so," or "How would Dad figure this deal?" It still, after ten years, takes some getting used to, and I probably never will completely. But I've got one more story about the man that is typical of him. The last story, maybe.

The winter of 1978-79 was the roughest I have ever seen. From early November until May we never saw bare ground, and for the first time in my life I fed every range horse I owned. For nearly four months. If I hadn't, they'd have starved, and we went into the winter with some nine sections of good grass. Our middle girl, Shelly, married and with four children, was teaching at the high school in Big Timber and made a forty-mile trip to town and back to the ranch every weekday. By early January, the school bus could operate only once in a while, so she used the Jeep pickup, staying in town when things were just too damn tough. A person using a Jeep, though, tends to get a little too brave and careless—sort of like a sharp-shod saddle horse.

So, one Friday afternoon, after being away from her husband, Bill, and her children since early in the week, Shell fired up the Jeep and headed for the ranch. The state had kept the blacktop passable, if you were careful, from Big Timber to Melville and west towards the Crazies for about eight miles. The county lane to our ranch turnoff quit the blacktop about halfway up it west of Melville, and from there to the place was better than two miles. The county had plowed the lane a time or two, but like all the country lanes, it had just drifted in again. They had other roads to plow, their outfit was broken down a lot from bucking the drifts, so finally, in utter frustration, they let us fend pretty much for ourselves. So the lane would have made a good team pulling a light bob-sleigh look to their hole card. Shell, though, made it to Melville, called and told Bill she was on the way home and got to the mouth of the lane in good shape. It looked bad, but she's bullheaded, like most of the family, so she put things in low low, turned into it, and, sure enough, in about three hundred yards she was stuck. Completely.

She still was all right, but then, for some knotheaded reason, she decided to walk to meet Bill and save some time. She was dressed warmly, she thought, so out she got and started trudging up the lane. She is pretty much of a runt, and there was around a foot of new snow on top of the crust which, when she broke through it—as she did every few steps—covered at least another foot of sandy snow. "Corn snow," I believe skiers call it. So the going was rank. To cap things, there was a low, nasty wind out of the west, which she had to quarter into, and the temperature was below zero. Not exactly top conditions for a walk! She told me about it later.

"I was cold and awfully tired when I got to the top of the Hart hill. The big drift on the crest broke the wind, and I thought I'd sit down and rest a while. It looked so comfortable and quiet. Then all of a sudden, Scrumper was beside me."

Dad and Shell had always been great pardners, especially when it came to chasing fillies, and she always called him Scrumper.

"I couldn't see him, really, I don't think," she went on, "but there he was."

"What did he do?" I asked.

"Well," a soft, loving smile spread across her face, "he swore at me, first. Told me, 'Dammit, get up, keep moving. If you stop, or sit down, you'll freeze.' Then he helped me up and walked ahead breaking trail. He didn't have a limp or a cane either, Dad, and his footprints made walking a lot easier. I don't know for sure whether he had hold of my hand. I guess maybe not, for he kept telling me to swing my arms across my chest. I had to stop a time or two, and he'd say 'Come on, Honey, keep moving. It's not far now.' So I did. When I saw Bill's lights ahead down by Dry Creek, Scrump said, 'Just a little more now. Stay with it,' but when Bill got to me, he wasn't there any more. Just like he'd come. But before he went, he chuckled—you know how he used to, Dad—and told me, 'That's my girl. We did it. God bless.' "

I believe her. Absolutely! That was Dad—taking care of those he loved. Alive or dead, by God, he'd do that.

=2=
THEY DON'T
COME IN BUNCHES

ADELBERT "DEL" WHITNEY was a State of Mainer. There were a lot of them in the early days, especially over on the Musselshell, and they had a lot to do with the making of Montana. When I got up into northern New England, I realized why so many of them had left, for it looked like a hard country to make a living in. At that, I wonder if once in a while they didn't regret moving. As the old-timer said, "If I had a summer home in hell and a winter home in Montana, I'd spend all my winters at my summer home." But when I was a kid, most everybody around was either from Maine, Missouri or Norway, with a sprinkling of Scots, Irish and Dutchmen.

I got to know Del pretty well, for he had some sort of dealings with Dad when I was a youngster. Later on I realized what a magnificent pair of people he and his wife were. I've always figured luck was mostly a matter of brains and using them, but once in a while somebody comes along who somehow is just plain lucky. Maybe he couldn't pour sand out of a boot, but if his horse picks up a badger hole, he's just as like as not to turn up a gold mine. Works the other way, too, and when a man runs into a plumb unlucky person, it's damn bad business to ride anywhere near him in a lightning storm.

Del Whitney was just that: unlucky. It showed up right away when he hit the territory. His stepfather, T. H. Gurney, who had gone west a year earlier, had a band of sheep on the Musselshell. He left them with a herder in the fall of '81 and went back to Maine for a visit. In the spring of '82 he and Del headed back, for Gurney had gotten word that his herder had committed suicide. They arrived in Miles Town March 9, took a stagecoach to Coulson and two days later left there with a freighter from White Sulphur Springs for the sheep camp. Right away the green kid from Maine sat on a prickly pear; possibly an omen of things to come.

On March 13—I'll bet it was a Friday—they hit their camp and found that their herder sure had killed himself. He'd done the business with a 45-caliber needle gun. A first class job, too. The little cabin was messy, to say the least; blood and brains were spattered all over the place and dishes. One cup was about half-full of dried blood. Of course, whoever had found the dead man had buried him, but that was all. As Del told it years later, "I was delegated to clean up the place. Father then told me to go to cooking while he hunted up the sheep." Sort of a rough greeting for an old Down East boy. Or for anyone else, for that matter.

He learned the business of sheepherding and dipping that summer. In the fall his stepfather went back home, and Del stayed with the band. He used to tell about how a camp of Flatheads came down the river after buffalo. The big kill had been in '80 and '81, so they didn't find any. After they'd left, he ran onto a pony which they'd evidently lost, and appropriated him. The horse could run a lick, and Del cleaned up. Every shearing crew and freight outfit had a pony they were eager to match, so he had, for about the first and last time, a run of luck. Finally, he sold the horse for $100 and a good team. The pony must have been a little rapid, for that was a pretty good stake.

It was over thirty miles to the post office in Martinsdale so he didn't get mail very often, and it was February of '83 when he learned that his stepfather, his mother and sister were on the way out and expected to be met in Big Timber.

The Northern Pacific had gotten that far west by then. Del went out to rustle up the horses, and as he brought them in, a wagon pulled up at the place. It was his folks. When Del hadn't met them at the railroad, they'd bought an outfit and come on over. His hair had grown down to his shoulders, and at first his mother didn't recognize him. It was great to see her again, and as he remembered, "It was like Heaven to sit down to her cooking. I wasn't strong on batching." Don't know if I'd been, either, if I'd had the start he did.

The summer of '83 he spent freighting wool to Big Timber. At that time the latter was one of the largest wool shipping points in the nation. Over 12 million pounds were brought in that season alone by bull teams and jerkline outfits. They back hauled freight for the ranchers or for Lewistown. In November he trailed 2,500 fat wethers to Billings to ship to Chicago. They cut through the Lake Basin on the way, and Del was impressed with its possibilities as a stock range. So, after the wethers were loaded in Billings—they were double loaded, the first time the method of two tiers to a stock car had ever been tried—he and a man named P. R. Lowell got a team and a camp outfit and headed back for the Basin. The snow was two feet deep when they got there, but they found a creek with a good spring, plenty of wood and shelter and set up camp. It got to 50 below, but Del claimed, "We had been living outdoors and were used to the cold." They were in a tent, mind you! Finally, the snow got too deep for their wagon, so they hewed out sled runners from a couple of small cottonwoods and went back to the Musselshell for the rest of the winter.

After the wool freighting season was done in '84, he and Lowell went into partnership and moved into the Basin, just over the ridge from the East Fork of the Sweet Grass. They spent that winter getting out logs from the divide for buildings. The summer of '85 was so wet they couldn't get to the timber for enough to finish the cabin, so they summered under wagon sheets stretched over the walls.

By '86 they had a going concern, and the next few years were so good that in '91 he went back and married a girl he'd grown up with in Maine.

Then that old tough luck showed up again. I don't know whatever became of Lowell, but the ranch was going fine and the Whitneys had three little boys when Del rode off for the roundup, leaving his wife to keep an eye on the place. I never learned exactly what year it was, but when he got home a couple of weeks or so later, two of the children were dead and the third dying. I think it was typhoid fever. Anyhow, Mrs. Whitney was so busy trying to cope with the sick children that she couldn't get word to anybody, and the neighbors were so damn far away that none had happened by. It was forty miles to the nearest doctor even if she'd been able to get word to him, so the poor woman, in spite of doing all she could, had to watch her children die. She had buried two all by herself by the time Del got home.

As if that wasn't aplenty, several years later the place burned down. Times were tough, they couldn't get enough together to make another start and were forced to sell out. With what they had left they bought a small outfit on Ten Mile north of Big Timber. After another few years of tough going, it was lost too. Once again they moved, each time with less than they'd had before. This time it was to Big Timber. Cars had gotten pretty common by then, so Del went into partnership with another man in the garage business. Once again luck was against him. As a last resort he filed for office and served as assessor of Sweet Grass County for three terms. Failing eyesight put an end to even this, but the Whitneys stayed in town living on what little they had salvaged and what odd jobs he could get. After his wife's death in 1944, he was completely alone, but with his sublime optimism and warm, gentle spirit he kept on, as lonely and as decent a man as I've ever known. At last,

completely blind, he had to go to the rest home. He died
there in February of 1957. Of the "infirmities of age." He
was ninety-two.

I was and still am a fervent admirer of Del and Lena
Whitney. I went to her funeral remembering what a lovely,
generous person she had been to an old gunsel kid; me. I
went to Del's funeral, too, and it bugged hell out of me that
I wasn't a pallbearer. I guess it shouldn't have for they
were all from his era. But, goddammit, I was one youngster, so to speak, who loved and appreciated Del.

I won't forget the last time I saw him. We were branding
at the home ranch. Mother had brought him out from
town, thinking he might enjoy it. When I saw who was
standing by the corral, I rode over to say howdy. "I can see
it all, Spike," he confided. "My nose and ears do it for me.
The smell of wood smoke, hot iron and burnt hair; the
grunts and banter of the wrestlers; the bawl of a calf being
thrown; the cry when the iron is laid on; the bellow of the
mother across the fence where she's working herself into a
rage. Ah, it's been a long time, but it's still the same, and I
love every bit of it. I only wish Lena could be here too. She
knew it so well!"

When we went to the house for dinner, I crossed the
living room to where he sat. "Del, you made it to green
grass again, and you sure look good."

"I did this time, Spike. I don't think I'll make another.
I'd like to, for you know as well as I that spring's the best
time of the year with the little calves, lambs and colts, and
the curlews singing. But would you mind crossing the room
again. I haven't heard the jingle of spurs for a long time.
Too long."

I did, walking a little heavily, I am afraid, while he
listened intently. A little wetness built up in the corners of
those sightless old eyes, and he knuckled them dry. "Thank
you," he told me quietly.

I brought him a plate of food and listened, as I so often

had done before, as he talked of times and people I hadn't
known. But never a word of complaint. About anything.
He'd taken it like it came. His breed did!

Maine has raised some fine people. Great people. But in
my books, none better than the Whitneys. The state
should be proud of them—tough, hardworking, yet gentle;
warm, generous, and uncomplaining. A State of Mainer all
the way, Del never outgrew his heritage. All his life he
spoke of rhubarb as "rhubub"!

One of the greatest ladies I've had the good fortune to
know in my lifetime was Mrs. Stanton Brannin up in Sweet
Grass canyon. She was a fabulous person. I never knew her
husband, but Gramp described him as "a redheaded, Irish,
old Apache fighter." Dad told of a time when quite a bunch
of deer hunters were up at Brannins'. They were talking
one evening of the chance of shooting one another by
mistake, when Mr. Brannin remarked, "Well, I'm goin'
huntin' tomorrow an' anybody takes a shot at me he damn
sure better make it a good one, for I'm goin' to be shootin'
back if I'm alive!" Evidently he was a good shot, for Dad
said nobody else went hunting the next day. I wish I'd
known him.

Mrs. Brannin was Spanish-American, her first name was
Guadalupe—a lovely name—and I have always regretted
that she died before I learned to speak Spanish. She came
from the Silver City region of New Mexico, and I think she
said that as a girl she knew Billy the Kid. Her daughter,
Babe, told me she remembered an old saddle in the family
which was supposed to have belonged to the outlaw.

The stories she used to tell, not only of growing up in
New Mexico Territory as a girl, but also of her twenty-five
years there as a wife and mother, were fascinating. One was
about some Apaches showing up at the ranch when she was
alone with the children. She convinced them that she was a
daughter of Victorio, who had cut a raw, red path through
the region on several occasions. They assured her that,

being who she was, she and the kids were perfectly safe. About then her son Dick appeared. He was a redhead, the only kid that wasn't real blackhaired, and the Indians were fixing to kill him. It must have been damn juicy, for if she admitted that this was her child, it was a fair bet that it would blow the Victorio story and likely they'd all be killed. But the indomitable young woman told them that Dick was the son of a neighbor and that she'd given her word to take care of him. So, if they killed him, they would disgrace the word of Victorio's daughter, and no true Apache could do that. It worked, for when they left, Dick was still in one piece.

In the spring of '95 the Brannin family left New Mexico. There were eleven children, a son-in-law, a granddaughter, a Mexican and Mr. and Mrs. Brannin. They all traveled in two covered wagons, a spring wagon, or ahorseback, for they had several hundred horses and burros with them and nearly a thousand Angora goats.

They arrived in Helena on June 6, 1896, "the day Billy Gay was hung," she used to say, short about half their goats and with a lot less horses. After several years spent in the mining areas, in 1903 the family moved over to the Sweet Grass, bringing along the few horses that were left of those which had made the long trip from the south and the remnants of the Angoras.

Mrs. Brannin was an extraordinary woman. Not only had she shepherded the outfit on the trip from New Mexico, but she was also very much the head of the ranch and family after her husband died. Could be that the boys didn't realize it, but she damn sure was. In her quiet, gracious way. No one could go by the place without eating—her feelings would be hurt if you did—and the time of day made no difference; you ate. If it was the middle of the night, she'd get up and cook a meal. You ate when you came to Brannins', by God. Yet I never saw her hurried or flustered. Just a lovely, big-hearted person.

I came into the ranch with a string of pack horses loaded with equipment from a Forest Service trail crew. I remember it real well, for among their stuff was a powder box about a quarter-full of loose dynamite. There was no way of padding it, so it rattled along behind us for two hours—two *long* hours—until we unloaded at Brannins. It was mid-afternoon, but we were fed a freshly cooked meal, and it was about dark when we left with our empty pack horses. The time had just flat slipped away, we had so much fun talking to Mrs. Brannin. I think she enjoyed it, too, for fresh news was pretty scarce way up there in the canyon.

Her personality permeated the outfit. She loved children, and I can't think of a finer place for a kid to grow up if he could put up with the steady joshing by the boys. Didn't seem to bother, at that, for I don't believe I've ever seen a happier bunch of youngsters than those that abounded around the place. Most of Mrs. Brannin's children took after her. Dick, the same whom the Apaches came close to killing, made me my first pair of chaps when I was just a button. Angora chaps, what else! Gus gave Barbara and me some goatskins when we lived in the Bennett saloon at Melville. The floors were splintery, and Barbie and Tack were real little. It sure saved their bacon, and we still have those skins forty-five years later.

Crawford, always known as Rube, was just a little guy when they started their trek north, had two bouts with typhoid on the way. It affected his speech from then on. He loved children and animals, and they responded to his childlike, gentle, generous heart. He had a damn fine sense of humor, too. The Brannins had a rodeo at the ranch. I won the calf roping; the first time I had ever done anything like that. Winning, I mean. Anyhow, during the dance which followed a sumptuous supper, somebody invited a few of us out for a shot from the jug he had cached. When I say jug, I mean just that—a big earthenware jug of moon from the still over the ridge, either ridge. Rube was in the group. The jug was alongside the house, and near it was an old boy who had either been sampling it or another like it, for he was passed out, snoring peacefully.

We each had a jolt. Then Rube got himself all fixed to say something. Pointing to the body and serious as a tree full of owls, he asked, "Why don't—they set—some traps— 'round him?"

His nephew, Buck Ward, told me about how, shortly after the Wards had gotten a radio, Rube was listening to it one evening while he smoked his pipe. Come to think of it, unless he was eating, I don't believe I ever saw him without a pipe in his mouth. Seems the announcer said something that annoyed him, so he puffed a big cloud of smoke at the radio. About then the announcer coughed. Tickled the hell out of Rube. He figured, by God, he'd caused it.

The Brannins' youngest daughter, Anita—though I have known her all my life as Babe—was cut out of exactly the same material as her mother. She married Bud Ward when she was only seventeen. He was an Englishman who had worked for my Uncle Tom Blakeman and Gramp. He was quite a horseman, but above all he bubbled with as much cheerfulness and wit as anyone I've ever known. He and his partner, Ernest Parker, ran a sawmill a couple of miles above the Brannin ranch for better than forty years. Two more dissimilar men never hooked together, for though Ernest was a hell of a good man, you'd never know it to listen to him. If it was a fine, bright day, he'd claim it was "a damn weather breeder." But he and Bud meshed perfectly.

Bud and Babe were quite a pair. All told she must have spent close to fifty years back in the hills, raised a big family, and I don't think I ever saw her without a smile. Bud used to tease her, "I found her in th' mountains. Had to run her down to get shoes on her, an' I've kept her in th' mountains ever since." He claimed, too, "Babe was raised on beans, billy goat an' biscuits."

Many a meal I ate with them when I was tending sheep camp in the head of the canyon when I was a kid. The road ended at their mill, so I picked up supplies and salt there. Babe was just like her mother when it came to feeding pilgrims. Much to my glee, for I'd been eating my own cooking in camp, and I sure wouldn't take any ribbons as a chef. Bud prided himself on his home brew, so if he had any ready I'd have a bottle before I ate. I remember going down to the water trough with him one day. There were six or eight bottles on its bottom under the icy water. "Have one," he said with a grin and handed me a church key. The cap came loose with an explosion like a rifle, and beer rained down on us. "Ah, 'tis a bit green. Handle it gently," and he passed me another.

Bud was a gregarious cuss and loved to celebrate when he got out of the hills, so it was pretty near a cinch he'd get something of a load on at Melville dances. As Gramp put it, "Bob Ward can get more fun out of a drink or two than most men can out of a bottle. He's always happy about it, to boot."

Bud took a different slant. "It's the altitude. Th' change in the altitude from th' mountains that does me in," he'd explain with a grin. "Not th' whiskey."

We were at a ranch for a party one night, and the booze got short so Rod Johnson and I went down to Melville to replenish the supply. Rod had some at his place, while I, just back from Arizona, had some tequila. Bud went along for the ride. When we got our liquor, he was fascinated by my tequila. "Is it guaranteed to?" he asked me, sizing it up carefully.

"Guaranteed to what?" I was puzzled.

"To kill you. Or something. That's what th' name says."

"Hell, no. It's pretty good, but sort of stout. Try some," I answered.

"I'll do that," and he uncorked the bottle, sniffed, and

took a good solid pull at it. He coughed a little and appraised the bottle judiciously.

"It looks like water, but tastes a bit like nails. Rusty nails." He had another snort and was sound asleep by the time we got back to the party.

Help was a little hard to come by at the mill, but one time Bud rustled up a man in Big Timber. His stay was short. The third morning after he arrived he came out of the bathroom flourishing Bud's toothbrush. "I've tried 'em all," he announced heartily, "but I like this'n th' best." The Wards took him to town that day—and got a new supply of toothbrushes.

They were fine neighbors. When I ran cattle in the head of Sweet Grass during the summers in the forties and fifties, it was quite a job to get them out in the fall. Barbara and I would camp in Eagle Park above the mill a couple of miles and ride up the canyon. I had a big steer horn bugle I used to call the cattle with when we fed in the winter, and I'd carry it on my saddle. As a rule the weather had gotten a little nippy, and hoping the stock had begun to think about winter, I'd start blowing it when we got up towards the forks. Lord, it sounded nice! The call would roll up between the rocky walls of the canyons, around the corners to the heads of the forks, die away, and then echo would sing back. Haunting; like a silver thread from away and away. Invariably a cow would bawl from one of the high glacial pockets. Then another and another. We'd head back, picking up the cattle along the canyon. When we hit camp, we'd shove them past, picket our ponies, eat and roll in. Whenever we'd wake during the night, we could hear stock stringing down through camp. That horn saved a lot of rough riding.

Of course, it didn't get them all, for there were always some individualists I had to make a trip or two for later. I did that alone, for the weather by then could be pretty juicy, particularly way up at the head. I'd either take a trailer as far up Dry Creek as I could, unload and drop over the ridge into Sweet Grass, or I'd pull up to Wards' and occasionally spend the night there if I was slow getting back from the high country. I was always welcomed and fed till my buttons nearly popped.

One go-round sticks in my mind. Barbara and I drove to Brannins'. She took the outfit home, and I went up the canyon ahorseback. Time I got past North Fork the snow was lying. I found all but two pairs of cows and calves on the way up and started them down country. Somebody afoot was ahead of me, but the tracks turned up South Fork. I knew there'd be no stock up there this late, so I kept on up Middle Fork. I finally ended in the "Hole" high on the side of the canyon, a place where cattle liked to hang out. Not a damn cow! It was getting chilly—it does at nine thousand feet or so in mid-November—and it was late. So I started back, blowing my horn every little while. My big mare was just as eager as I was to get the hell out of there, and she came down the mountain like a goat, at a long, swinging walk. We hit the trail in the bottom, and she broke into a jog. She was doing it when we came around the corner into the park at the mouth of South Fork, and I looked into the barrels of two rifles, a tense man behind each one. It was still fairly light, thank God, and they relaxed as they saw me.

"Christ, Spike, I'm glad it's you," said one.

"Sure doesn't look like it," I told him. "Just what the hell's going on?"

"We been after a buck up South Fork, an' about th' time we got here we heard some damn thing buglin' up th' canyon. Comin' towards us. Somethin' big, from th' sound an' we didn't know what in th' hell to expect. Sorry we spooked you."

"I'd say it was about six of one and half a dozen of the other, wouldn't you?"

I knew them both, but they were camped in the timber

off Eagle Park, and I'd missed it. Anyhow, we travelled together until my mare left them. I picked up what cattle I'd passed and pushed them down to the mill. There wasn't anybody around. Come to think of it, there hadn't been when I'd gone up, either, but I put my horse in the barn, started a fire in the cookstove and waited. Nobody showed, so I rustled myself some supper, washed the dishes, hunted up a bed and turned in.

I was still alone in the morning. I ate, took care of the utensils, filled the woodbox, and tended my mare. When she'd finished her grain, I saddled up, cleaned the stall and just on a hunch went back up the canyon. Sure enough, before I got to the park, I found the last two pair of cows and calves that made my count right. The sound of my horn must have jarred them loose from wherever they'd been hanging out. By afternoon, I was home with all my cattle.

I ran across Bud a month or so later, told him I'd spent the night and thanked him. "Glad you did," he assured me. "I knew who it was."

"How come?" for I hadn't left a note or anything.

"By th' stall. You cleaned it. These other bostards never'll do thot."

The Ward children were a nice bunch. Bob, Jr., "Buck," was a dandy. He ran the calf chutes at the Melville Rodeo for us. A good man and a happy, uninhibited son of a gun. Always up to some damn thing. He and his wife, Jean, lived back in the timber from his folks' place. One night, when she was putting their little kids to bed, Buck donned a wolf mask, scratched on the window, peered in, and absolute chaos resulted. The poor kids had been brought up on wolf stories, and here was one glaring in the window. It scared the liver and lights out of them. It was months before they got cooled down. And Jean even longer, only she wasn't scared. Buck was in damn bad odor for quite a while but completely unabashed. Except when he was around Jean, that is. He's a minister now, a popping good one, and his sunny disposition hasn't changed a lick.

Mary Jane, the youngest Ward girl, was about Barbie's age, and the two were great friends. She took after both her mother and dad, so naturally was a great favorite of mine. She, too, was a cheerful, irrepressible individual. I was at the mill one day talking to Bud when she brought some cattle off the mountain. They were giving her lots of trouble, she was by herself, so her language got more and more choice. When the cattle broke out of the timber into the bottom and we came into sight, she toned it down. As she rode past us, Bud said seriously, "Mary Jane, I'm ashamed of you. Such profanity!"

I had been sympathizing with her, for I'd brought my cattle down that cut-over damn hill a number of times. It was bad. Nonetheless, I looked shocked and put in my two bits worth. "For shame. A nice girl like you, too."

She stopped her horse and sat sizing us up, bright-eyed as a chickadee. "Huh. I don't know anybody in Sweet Grass county who has less business talking about swearing than either of you two reprobates," and she headed after the cattle. We grinned at one another, for she had us there.

Bud is gone, and Babe lives in Big Timber now. I miss them every time I go by their old place. It's so damn empty. Bud should be sitting on the stump by the yard. Puffing his pipe and ready with some cheerful remark. The house isn't itself either without Babe standing on the porch. Smiling and asking, "Have you got time to come in and eat?" A grand pair of people!

The winter of '17 about cleaned Dad out of cattle, and that's how I came to know Charley Bair. He was one of the giants of the sheep business in the United States. He had holdings at the home ranch near Martinsdale on the head of the Musselshell, and God only knows how much land he leased on the Crow Reservation. A man who grew up along the route used by Bair sheep from Billings to the Crow, told me that he had seen bands pass almost without a break for two days and more hand running. I have no idea how many he owned, but the number must have been enormous. The story goes that, when he was back in Boston for some wool doings or other which included a banquet, one of his tablemates, a staid, proper New Englander, mentioned that he, too, was a sheepman. "Sure enough," said the Montanan. "How many do you have?"

"A hundred and fifty," the man answered proudly.

"Hell," Charley snorted, "I've got more *sheepdogs* than that!"

Which was probably absolutely true, for in 1910 he sent east the largest single shipment of wool ever to leave Montana on a railroad. Forty-four carloads!

Sheep weren't all the livestock he owned. He ran cattle too. In '19, knowing what a licking Dad had taken, Charley staked him to a string of cattle on shares. He would drop by our place every now and then, he and Jim Bowman, his cow foreman, and I remember him well. Friendly, optimistic and generous as all hell. I liked him, even though I was just a kid.

The winter of '19 was worse than '17, though it's a cinch somebody will disagree with me. Dad had carried over quite a bit of hay, seeing as he hadn't had much of anything to feed it to in '18. Besides, he was a believer in Gramp's rockbound rule, "Never, ever, sell hay. Lend it, sure, but don't sell it." When the hay finally ran out, Pop started hauling cotton cake from Big Timber. It was new in the country then, but the cattle sure liked it. So did I, for I had to give it a try. Tasted real good, though the strands of burlap in it were a bother to get separated after you chewed it for a while. If you didn't get them all out, what the hell; gunny sack never hurt anybody. Depending on how rough things were, Dad drove four or six horse teams on the wagon. A bob sleigh was no good on account of the damn wind down towards the Yellowstone baring the road. He used to hang a bell or two on the pole. Said the sound sort of kept him company. We sure weren't living high off the hog that winter, so Pop carried his old 32-40 with him. He could use it and would get back from town with a dozen or more jackrabbits piled on the sacks of cake. Every damn one shot through the head. I remember it real well, for I helped skin them. We wintered mostly on oatmeal, hot cakes and rabbit. Mom had a way of cooking the latter that she called "jugged hare," and they were pretty good. I guess. I still swear by oatmeal and hotcakes, but I've never knowingly eaten jackrabbit, by God, since.

The cattle wintered in good shape. We summered them in the basin south of Wolf Butte, and they came out rolling fat, much to Dad's and Charley's satisfaction.

That fall Dad took a relay string to Billings to its fourth annual fair. I got to go along with my mare, Panama. It looked like there were more redskins in town than there were whites. They impressed me more even than the first coconut I had ever seen—I think it was in Yegen Brothers—and I couldn't figure what in hell it was. Charley must have heard me questioning Mother about it, for be damned if he didn't turn up with one for me. Even showed me how to get into it when I was plumb stymied.

Dad won the relay. A close thing, for the cinch of his relay saddle was cut part way through before the race the last day—pretty much an old-timey race horse trick, and I could put a name to the man who did it, right today. The cinch broke on his final horse, but, being Dad, be damned if he wasn't able to stay aboard and finish the race carrying his saddle.

Charley had a bet on Pop with Walter Hill, the son of Jim Hill. The latter built the Great Northern Railroad and was probably as intensely hated by the stockmen of Montana as any man who ever lived. Anyhow, Charley won the bet—a new hat, a new pair of boots and a new suit—and he got the best Billings had to offer.

Maybe that's why he collared Panama and me and took us down to the Crow camp. I guess I better quote from a late September, 1919, issue of the *Big Timber Pioneer* as to what transpired:

> Paul L. Van Cleve III of Melville, only six-years old but a comer, won second money in an Indian boy race at the Billings Fair. Before the race, he was taken to the Indian village by Charley Bair and introduced to the Indian boys as "one open to all comers." He was then introduced to the grandstand by starter Keown as, "Paul L. Van Cleve III, who is going to clean Indian house." He did not do it but came near it, an Indian boy of thirteen, dressed only in a breech clout, dubbed by the announcer as "September Morn," winning first money."

Actually I was pretty near seven. In a couple of months, anyhow. I don't remember feeling real badly about losing, probably because I was so overwhelmed by all the excitement and foofaraw. Besides, the Indians were friendly as could be both before and after the race. Especially after. Matter of fact, September Morn and I decided we'd always be compadres. I wonder what ever became of him. Might be that Charley figured I needed cheering up, though, for that evening he gave me a double eagle—the first gold coin I had ever seen—and told me, very seriously, to take good care of it. I have, for sixty-three years. I must admit that I was tempted a time or two when Barbara and I were scratching gravel, but I didn't succumb. I look at it once in a while, and every time I do I think of one hell of a good man who helped my dad through a tough time and was awfully nice to a little boy. Even if he did get him outrun.

3
COWBOYBOOTS

USUALLY WHEN I FISH around home, I wear Levis and felt-soled shoes, for I always get wet—fall in or something. Then I change into dry pants and boots when I get back to the car. This day I'd had a few guests with me, it was nice and warm and so when we came off the creek, I decided just to wear my fishing clothes on the drive home to the ranch. As I put the extras into the trunk, a woman who had been along cautioned, "Don't forget your cowboyboots."

She was a guest of long-standing, so as I started to answer, she grinned shamefacedly and added, "I knew when I opened my damn mouth I'd made a mistake, Spike. Don't forget your *boots.*"

She savvied! To me, and I think to most ranchers, *boots* mean one thing: boots. Sure, there are irrigating boots, gum boots, English boots, jodhpur boots, hunting boots and so on. But never, ever, cowboyboots. Just boots.

It beats the hell out of me how many people wear boots nowadays. From the looks of a lot of them, too, I sort of suspect that a high percentage must speak of them as *cowboyboots.* Clerks, barbers, mechanics, bartenders, lawyers, bankers, and the law, for sure, to name a few I've seen. I have my suspicions, too, about those big old boys in shiny new pickups with a rifle, buggy whip and carefully coiled catch rope hanging on the gun rack across the back window and a bale or two of hay in the box.

Boots have been a part of my life. Of course, I never

31

owned a pair until I was pretty good-sized, but I sure was
around a lot of them as far back as I can remember. The
first was that high-topped (pretty nearly to the knee),
round-toed, almost like a shoe style with the long, leather,
"hound ear" pulls hanging down on each side of the tops
that the old-timers favored. Then the style changed, and I
think that rodeo, which showed up in this region about
then, had a lot to do with it. Narrower, square toes began to
show up, heels got higher and 'way undershot and the tops
shorter. A few bronc riders, though they took to the
trimmer feet, still stayed with the high tops. I remember
Yak Canutt at the Bozeman Roundup in '21. His boots
came just short of his knees, he tucked his britches in them
and about half the time he rode his stock without using
chaps.

I've heard it told that the heels were set under to make it
easier to r'ar back on a rope afoot, but I have my doubts.
Hell, a man wasn't afoot except in the corral, and every
good set of ranch pens had a snubbing post. Maybe they did
help forefooting broncs, though, at that, but it would have
been about the only place. Personally, I think they got
popular because they looked real good, and of course, they
held a stirrup well. "Beartrap" saddle trees were in vogue
then and the narrowest of oxbow stirrups, and I wonder
now why, particularly with the caliber bucking horses in
those days, there weren't more broken backs and bad
hangups at the shows. Luck, ability or maybe everybody
just got used to the damn riggings.

It was a lot different with the lady bronc riders. Most of
those I knew, barring the Greenough girls, rode with
hobbled stirrups, and I be damned if I don't believe that's
why so many of them got killed or badly crippled in the
arena. It was a fine event, and some of the ladies, Fanny
Sperry Steele especially, took their stock just as they came.
It's a shame the event has pretty well died out by now.
Those girls were damn sure hands!

The first pair of boots I ever owned I got from Blucher in Olathe, Kansas, and I don't believe I ever had a better pair; French calf that wore like iron, five rows of stitching on the tops and made to measure—$21.50 for the works. Besides, when the feet wore out (which damn sure took some doing), Blucher'd refoot the tops for $7.50! They could do it three times or so, at least until the throat of the top got too narrow, for they had to take it in a little each time.

Those Bluchers were almost new when we took a pack trip over the divide into the head of Sweet Grass. I don't wear boots in the high country where I am liable to do much walking—boots are for riding. Even then they were too high-priced to wear out afoot or get all scarred up, for money was damn hard to come by back in those times. But, though I was wearing shoes, I took my boots along. Don't know why, unless it was just to admire them; or maybe I figured on wearing them around the fire in the evenings to impress the girls. Anyhow, I put them on a real gentle pack horse—we'd put the eggs on him, too, he was so quiet. And wouldn't you know it, about two-thirds of the way down the yon side of the hill the damned old fool tripped or something and rolled plumb to the bottom.

Dad was on the fight about the eggs, for all of them—enough to feed about thirty people for three mornings—weren't just broken, they were absolutely mashed. Where the old pony wasn't red, he was egg. But it was my boots that worried me. There was egg on them, egg in them from hell to breakfast. I wiped them off with grass, washed them carefully in the creek and tied them to my saddle upside down. When we made camp, I talked the cook out of some bacon grease, worked them over with it a couple of times a day, and they finally turned out fine. Still, since then, when I take boots anywhere, I wear them. Incidentally, we left the old pony in the grassy pocket where he'd lit (there was water there) and when we picked him up on our next trip over, he was in good shape, barring a lot of white patches of hair.

Some horseback man must have once said—and all of them since then have believed it—that riders have smaller feet than shoe and sock people. I guess it's that belief, plus pure vanity, but I never knew a man to measure himself for a pair of boots that he didn't get them a little too snug. At least, not until he was about fifty years old and had weathered some miserable experiences. Matter of fact, you could pretty nearly always tell a man with tight boots, even if you didn't see him walk, by the diamond-shaped pieces of leather on his insteps; it was a cinch that he'd had to slit the instep to ease the pressure and had gotten a nice patch put on over the cut.

The medieval instrument of torture called "the boot" must have been named by a man who sure savvied, but I doubt that it was a damn bit worse than tight boots, afoot or ahorseback, now. A man can feel each heartbeat agonizingly from the top of his head to his toes! I complicated matters once, only once, by putting neatsfoot oil on a new pair I'd measured myself for. The boot on the sunny side of my horse drew down something fierce. The one on the shady side wasn't so bad, but hell, a man going somewhere can't go in the opposite direction half the time or he just as well not start in the first place. Behind the drag where I could swap back and forth, I could stand them, but they never got what I'd call wearable until I had the local cobbler work them over with pretty patches.

Reminds me of a pair Blucher made Dad. He'd measured himself, and while they were fine boots, kangaroo with pretty redbird insets in the tops, they were tight and got tighter. They were so good-looking, though, that he just, by God, couldn't bring himself to have the insteps slit. Things came to a head when he wore them back East on a trip to drum up dude business, figuring maybe they'd limber up if he did a lot of walking in them. They didn't, so he bought a pair of shoes to use except when he was with prospective guests. Those times he just crippled around and, as they say, kept a stiff upper lip.

A man at some dinner party admired them and insisted on trying them on but be damned if he could get them off after he'd worn them a while. Everybody had a try at getting him loose from them with no luck. So, the upshot was that the guy had to wear them home and even to bed, though I don't imagine he slept a hell of a lot. They finally got them off him the next day, and I cringe to think what his feet felt like when they finally came free. They must have immediately resembled a couple of red toy balloons —in the process of being blown up! In any event, the guy never came to the ranch, and Dad made me a present of the boots soon as he got home.

They fit me fine, and I used them for a long time. First for dress and after they got worn for work. When the feet finally wore out, Blucher was out of business. But we were wintering down on the Santa Cruz, I knew a good boot maker in Tucson, and he refooted them for me. It was just before the war, and I supplied the leather. Where or how I fell heir to it, I don't remember, but it was the damndest stuff! One batch was a pretty blue. I had plenty of it, so since it was a little gaudy for a man to use in those days, I had a pair of boots made from it for Barbara. The other was a fake alligator which I used for the feet of my redbird tops. The man measured us—I'd learned better than to trust myself by then—and the boots fit fine. Trouble was, the damn blue leather never seemed to quit stretching. So Barbara, after going from silk socks to wool ones in about three months time, swapped them to Jack Avery, and they fit him to a tee. I haven't seen Jack since that spring, but my bet is that in his turn after a few months, he had to deal them to some really big-footed old boy.

The leather in mine was just the opposite. They shrank and kept on shrinking, seemed like. They fit to beat the band when I got them, but a couple of weeks later when I got gussied up for a dance, they seemed tighter. They damn sure were, too, and I ended up dancing in my sock feet. I used them as often as I could stand it and even got the insteps slit, but be damned if I don't think the leather in the feet got the patches to drawing, too. I bowed my neck and stayed with it, though, and it was only when I was getting ready for a party and heard one of the kids tell Barbara, "Hey, Mom, Dad's putting on his getting drunk boots for tonight," that I backed off and really sized things up. Come to think of it, I hadn't stayed sober a single time when I was wearing the damn things. Couldn't. Even drunk they were bad, until I had passed a certain stage. And it was taking more and more booze to reach that stage. So I quit trying to wear them, and I have a hunch that is part of the reason Barbara and I are still married. Sure helped anyhow.

They were good boots, though, so I saved them and put them way back in the closet where nobody could give them to a clothing drive or something and when Bobby was a teenager, gave them to her. She used them for several years, and when they began to bother, Shelly, who was even more of a runt than Barby, fell heir to them. I guess she wore them out. Finally. But through all those years and even now they are remembered as, "Daddy's getting drunk boots." Barbara, though, called and still calls them, "Daddy's G D boots," and I don't think she means "getting drunk!"

I've noticed that as people get older a lot of them seem to quit the narrow-toed, skintight boots. Old-timey round toes and roomy feet take their place, and men that have walked like they were on eggs for the past thirty years get around afoot in a pretty limber fashion. I have an idea that the change is a result of bitter experience, plus the fact that as a man gets a little long in the tooth vanity gets backed off by comfort. Al Jenkins in Billings where Dad got his boots for twenty-odd years told me that when Pop ordered his last pair he wanted them made without the stiff lining in the toes. Wanted them completely soft. Al got them made that way for him, and I couldn't help grinning when I heard Dad proudly tell people, "Best damn boots I ever owned."

Personally, I've always favored high tops and tucked my britches in them, for they are pretty near as good as leggin's for protection in this north country. Besides, if a man wearing low top boots is running fillies where it's sandy or there's lots of gravel, come night he'll have both tops full of rocks. In brush, he'll pick up enough free firewood to cook supper, too.

We had a young fellow, an Easterner, working for us who ran his right boot heel over something terrible. He was grumbling around about it one day when Riley Doore told him, serious as a preacher, "Hell, son, if your folks had taught you to turn to the left oftener you wouldn't do that to your heel." The youngster believed him, so every time one of us saw him start to the right, we'd get him to whoa up, and he'd dutifully head the other way.

Another big, pink-whiskered kid who was our choreboy one summer had a sorry old pair he delighted in wearing. They looked like hell—lowered the tone of the whole outfit. Originally, they must have been made by one of those companies whose boots, as Riley put it, "looked like old buffalo shells after they'd been out in a rain or two," and on top of that they were plumb worn out. The kid claimed they were too tight—though I don't see how they could have possibly pinched his feet any worse than a gunny sack would have—and asked what he could do to stretch them some. I said, "Sure, fill the feet with oats at night, pour some water on the grain, and they'd be loosened up by morning." They sure were—looked like a charge of powder had gone off in each one! The boy complained bitterly, but we finally convinced him that he'd just used too damn much water. All's well that ends well, for Bob Langston sold him an old pair he had. They looked a lot better at least for a while, and I never heard the kid fuss about their fit, either.

When Irby Mundy was rodeoing back in the twenties and early thirties, roping on his bob-tailed horse, he peddled Hyer boots as a sideline. Had a lot of samples with him, and at one of the shows somebody stole the whole works. Mundy was pretty warlike about it but seemed to get some consolation from the fact that all his samples were for the same foot. "You fellers let me know," he said, "if you happen to run across some one-legged sonofabitch who's wearing a new boot, an' I'll be obliged."

Nobody ever did as far as I've ever heard, but I'll bet whoever got the boots was a disgusted booger when he looked them over. Probably about as bad as Irby.

Every now and then when I see an ad somewhere that mentions cowboyboots, I'll grin. Then I'll cut it out and mail it back East to the girl that was fishing with me. Not always, but often enough to make it worthwhile, I'll get an answer, "You can go plumb to hell, Spike!"

4
IVAN

IVAN WASN'T MUCH of a horse, as horses go. A catch colt out of one of our work mares by a good Hancock stud we'd raised. He hit the ground all feet, joints and head, but his mother milked like a Holstein, so he grew, Lord, how he grew. But he never seemed to catch up with his feet and head. As a weanling, he was goopy, clumsy, damn sure hadn't been up towards the lead when smarts were given out and was always into some sort of jackpot. If there was any possible way he could get into trouble, he did. Matter of fact, if there was *no* way he could get into a bind, he'd find one, somehow. Looked like he had a gift for it, but he always came out of his storms with pennons snapping.

I remember once, when he was an overgrown three-year-old, finding him down and lying on his head. Groaning sepulchrally. What had happened and when, I had no idea, but I got a rope on his front feet. Then, by dint of much spurring, I got my saddle horse—who was sure this black thing was either dying or about to attack—up to where I could take a couple of dallies and roll the colt over. Up he got, shook himself so hard that he came within an ace of falling down again and thundered away proudly, tail in the air. Like he'd done it all by himself.

He grew into a big, awkward horse. Easygoing, curious, friendly and trouble-prone. We'd have broken him for harness, but about that time tractors had pretty much replaced teams here in the Melville country, and it was

getting hard to hire hands who were worth a damn as
skinners anyway, so we broke him to ride. His attitude, first
time he was saddled, was, "What's this all about?" Then
came a burst of excitement—clumsy, uncoordinated but
loud attempts at bucking—that culminated in a spectacular fall. When he got up, there was much nose-blowing and
craning around to smell of the stirrups. Then he pricked his
ears, walked over to where we'd been watching and sighed
philosophically. If he'd been able to talk, he couldn't have
said it plainer: "The thing's still there, so to hell with it.
But could any of you people help me?"

The horsebreaker, Joe, picked up the macarty rope,
stepped aboard, and the black colt moved off without a
bobble. Joe grinned, "I'm goin' to call him Ivan. Damn, if
he don't remind me of one of them big, rough Roosians I've
seen in pictures. All is missing is the beard."

That was it. He was named and broken, all the breaking
it ever took. He made a dude horse. Damn sure not one of
the best, but he was handy to have around. Fine for a big,
green man who wouldn't get out of much more than a walk
anyhow or a little kid who just sort of went along for the
trip, with Ivan, to all practical purposes, empty and plumb
free. Then, too, if I had trouble with some big, smart
teenager who figured that the only way to ride was at a
hard run, Ivan was just the ticket. I'd change the offender
onto him, and it usually took about one gallop, two if the
kid was real tough, for a rider to get a bellyful of his more
rapid gaits. I've heard bullfight afficionados speak of a
matador, "setting his feet like they were grown to Mother
Earth." That would describe Ivan's lope. My God, it was
rough! His trot was a little better, I think: four separate
jolts instead of the two spine-shattering shocks that
characterized his run.

He made a packhorse, too, though I never put eggs or
anything else on him that was, by any stretch of the
imagination, breakable. But for salt he was perfect, and,

big as he was, he carried as high as three hundred pounds of
it many's the time, though two hundred is usually considered tops for a pack animal. To hear him coming up the
trail when he had a load of utensils and metal dinnerware
was an experience. No matter how carefully the load had
been packed and padded, it didn't take long for him to jar
things loose. Reminded me of the man who was listening to
a carillon. He was enjoying it and mentioned the fact to an
old fellow who happened by. After about three tries at
understanding, the old guy reared back and bellowed,
"Can't hear a thing you're saying on account of them
goddam bells!"

Ivan, though, came along in the midst of his cacophony
seemingly oblivious to it. The guy leading him and the
horse he was being led from sure weren't—nor anybody else
within a half-mile, even upwind.

Nope, Ivan wasn't much of a horse. Clumsy, rough and
bumbling. But he was honest, friendly and dependable. A
good guy, the son of a gun.

Three years ago in early November, a neighbor called.
He'd been hunting some missing cattle, and while having a
look through our upper pasture on Otter Creek, he'd run
into a crippled horse on the bottom of the south fork. A
black gelding with the VC on him. In damn bad shape. He
hadn't had a gun, and things up there were tough, he said. I
thanked him and spent the night worrying about what
horse it could be.

The next day I saddled up, tied a halter to the fork and
slipped my 25-35 into the scabbard. Just in case. It was a
mean day. Cold and with the feel of a storm. Of course,
we'd already had snow a couple of times, but the wind had
put most of it in the coulees, so we made good time. When
we dropped into the creek at the forks, I was surprised how
much snow there was in the bottom. Where the wind
couldn't reach it. Might help, though, if it came to doing
some tracking. But there weren't any tracks, even when we

got up as far as the middle crossing. Looked like everything
had pulled out for the lower country.

From the crossing to the reserve line a mile or so above,
the canyon narrows. Its south side, where it isn't rim, is
small timber so thick a snake'd have trouble getting
through it. The north side is a steep, open hill with
scattered trees. The bottom had had a stand of sizeable
spruce, but a cloudburst a couple of years before had cut it
all to hell, so our progress was sort of steady by jerks.
Washouts, boulders, piled timber—all covered with about
eighteen inches of snow. I was riding Sam Hill, my old rope
horse and one of the best I ever threw a leg over. We were
both getting long in the tooth, but he'd always sported a
vivid imagination, so I had to spank his seat a time or two
before he quit boogering at things. Or maybe it was that he
just got so busy trying to find solid places for his feet that
he forgot about the spooking.

The footing went from bad to worse, so I got down and
led him as we picked our way along. Finally, it got so
impossible we pulled out of the bottom on the north side
and worked gingerly along the slope a hundred yards or so
above the creek. Why in hell a horse would stay in country
like that, I couldn't figure, but Gene had said it was on the
south fork. Besides, I knew there were a few pockets up
above, so we kept on.

We passed one pocket, empty, on the bottom below us.
Then another, and I was about ready to give it up when
Sammy cocked his ears. I studied what I could see below
and ahead. Nothing, so we went on, slipping and sliding.
We'd made about fifty yards when Sam set the brakes and
almost jerked me over backwards. I turned to cuss him, but
he had no time for me, for, ears up, eyes showing a little
white, he was staring intently down the slope. Then he
whistled and moved up against me for reassurance. And
what I had taken for a charred stob at the edge of the
timber rimming a little half-moon of snow below me,

moved. No sound, but when the head lifted, I saw that it was, sure enough, a horse.

Just to be on the safe side, I put the halter on Sam and tied him solidly to a tree before I got the rifle and started down the hill. Most of the time, I was on the back of my neck, so I got to the bottom a damn sight quicker than I'd figured on. Standing up and brushing off the snow, I squinted through the shadows to where the animal stood— it hadn't moved during my gaudy descent. It was Ivan all right. He was so poor that I wouldn't have known him if it hadn't been for that head. "What the hell have you done now?" I asked, and at the sound of my voice his ears came up, and he whickered softly. He'd known me all his life. I was there, finally. Everything was going to be all right.

The look, the smell, of his off front foot knotted my stomach. No wonder Sam had been leary. Wire had cut deep into the ankle joint, it had gotten infected and was so swollen I could barely see the hoof. As I petted the rough hair on the poor, thin neck, I tried to figure what happened. Grazing with his friends along the steep side hill above, he'd probably slipped in the snow on account of that bad foot and had ended down in the bottom before he could stand up. Then he hadn't been able to climb back, and the washed out creek, ten feet of sheer drop, had stopped him on the other side. The horses had gone on and left him. How long ago it had happened, I couldn't tell, for the marks of his slide were hidden under the snows of the past weeks. Only the tracks of Gene's horse showed clear on the slope. It'd been a while, for the pocket was churned from his pawing and dirt showed where he'd dug to the grass roots. Bloodstains, too—he couldn't put any weight on the bad foot so he'd stood on the other and pawed with the crippled one. Godamighty! What few trees there were had been skinned, needles and bark, as high as he had been able to reach. I shivered. Alone—and horses love company; hungry; always in the shadows—the sunlight hadn't reached the pocket for the last month and wouldn't until some time in March; cold—it had been below zero several times already; crippled. . . .

There were other things I read, too. Bird droppings on the bony back and down the ribs. Magpies. But they hadn't started working on him yet. He must have still been able to shiver his hide and switch his tail to run them off. But there'd come a time when he'd just get too tired, and they'd start pecking on him as he stood. Coyote tracks around the edge of the pocket. I could see where they'd sat and watched. *They* had plenty of time. Eventually, he'd be too weak to stand. When he was down, they'd move in. They'd begin on the flanks, the tender spots, where they could work in to the intestines. While he was still alive.

"Old son," I told him, as I scratched the matted forelock and the cracked nose poked at my pocket. "Damn me to hell, but I haven't a thing."

Not a thing. Not even one lousy pellet! Ordinarily, I never ride through the horse range in the winter without some oats tied back of the cantleboard, or at least a pocketful of pellets. I'd been so anxious to get away in the morning that I'd just flat forgotten. Damn and damn and double damn! My eyes filled as he quit hunting, raised that rough old head and those tired eyes asked, plain as plain, "Come on now. Where is it? Please."

Nothing. Not even one single oat kernel to put a good taste in his mouth. Well, there was one thing I could give. Yes.

I patted him gently between the eyes and walked off about twenty feet, so I could do a clean job. He tried to follow me but couldn't cut it. But his ears came up again—maybe I was getting something after all. I worked the lever of the rifle, blinked my eyes clear and put the bullet exactly where it belonged. He dropped like a stone and stiffened. Walking over I knelt down to make sure he was dead. My hand on the chest back of the foreleg felt the big heart slow and then stop, as the body went limp and the snow under his head reddened. The hot smell of it reached Sam up the hill, and he snorted. It must have drifted up the creek, too, for a magpie slid in to watch from a limb. "You can have him now, you bastards," I told him. "You and your yellow-eyed friends."

One last pat on the warm shoulder. I shut the lid of the already clouding eye and was surprised to see how much grey showed above it. Damn, I'd forgotten how old he was. Well, didn't matter now.

Sam made a perfect ass of himself when I got up to him. "Cut it out, you old fool," I snarled, got aboard and spanked him, harder I guess than I really should have. He'd known what went on and, come to think of it, I'd very probably have to do the same to him some day. Might be that I'm silly, and a poor businessman to boot, but when a horse gets too old or stove up to use, I won't sell him for a canner. Even at today's high prices for killer horses. No, by God! When a horse who has worked honestly for me all his life gets to the point where he can't do it anymore, he gets pensioned off. Retired on the best range I've got. No being crowded into a truck with a bunch of strangers, taken away from the country he knows and whipped up a chute in a slaughterhouse smelling of horse blood. Instead, he lives the life of Riley until the time comes when I see he can't winter. Then there is a bait of grain and a quick bullet. At home, where he belongs. It's a dirty job. The worst, I think, that there is on a ranch. For a man who cares, anyhow. I always do it myself—I'm afraid somebody else might mess it up. A lady or a gentleman, and horses sure are, deserves the courtesy of a clean, decent death. As an old Mexican once told me, "To die with dignity. I ask no more."

During our trip home the snow and the northeast wind started. By the time Sammy was tended to and I headed for the house, a sure enough blizzard was pouring it on. The first bad storm of the rankest winter I have ever been through.

I was slow getting to sleep that night. As the wind whined around the corners of the house, whipping the snow against the windows, I could imagine what it must be like in that pocket up against the mountain. And I was glad, thanks to a good neighbor, Gene Langhus, for telling me about him and to Sam for finding him, that old Ivan wasn't hurting. Or alone. Or cold. Or hungry, out there in the storm. No. But, dammit, I should have had something to feed him first.

═══5═══
WHEN I WAS A BUTTON

HILMAN GUNDERSON was one of the first children born in the Settlement and sort of a fixture. His dad, Julius, was a dyed-in-the-wool horseman and owned some salty teams, but Hilman must have had a liking for machinery. Anyhow, he put together a rig and when I was growing up, did most of the threshing around the neighborhood. We kids sure figured he was something. To see him piloting his big steam engine down the road, the separator behind it, was kind of like watching God. The rumble, the smoke, the hiss of steam and the quavering note of the whistle he always blew when he passed the schoolhouse gave us the shivers. Why, our respect even slopped over to his helper who followed along behind with the water wagon.

I'm sure Hilman enjoyed our adulation, for even though he wasn't much of a talker, he always took time to show us all about the engine and machine and to answer our innumerable questions. He had a habit that tickled us, too. He'd start all his answers with either a "Hmmm, God yah" or "Hmmm, God no." Always. He really enjoyed traveling around with his rig. Got all the news and sure ate well, for every ranch wife was on her mettle when it came to feeding the threshing crew. He was a bachelor but never seemed as keen on sparking the local schoolma'ams as most of the other young bucks. Possibly because he was a little older, but mostly, I think, because he was sort of shy. There was one time he got interested in a girl though, and it sure sticks in my mind.

47

Lars Wesland and his wife were starving out about then on their homestead north of Big Timber Creek. He was a good man, but hell, the place wouldn't have run more than maybe one milk cow the year around, no matter how hard he worked. His wife was a real nice lady, French-Canadian I think. At least her name was La Bree, or that's how it sounded. How the hell it came about, a Norsk marrying a Canuck, I have no idea. She wouldn't have taken many ribbons maybe, but she was happy, friendly, optimistic—which damn sure took some doing where they lived—and my bet is that she made their life as pleasant as it possibly could have been.

She had a sister who visited them for a while. She was a lot like Mrs. Wesland. Younger, happy, chunky, talkative, and her name fit her to a tee—Toots. Anyhow, Hilman took a shine to her and went to doing a little courting. It's a cinch that it wasn't a real impetuous deal, but they were so different they sort of fit. He tall, she short; he spare, she a little hefty; he quiet, and she anything but. Things seemed to rock along in good shape, for he acted like he was really gone on her.

I remember a party at our place during Christmastime. The two were there, and during the festivities we happened to have a game of Ruth and Jacob—or rather, as pronounced in the Settlement, "Root and Yacob." Happened that Hilman got to be "it," so, as the ring went around, some joker slipped out, got into a kimono and apron and a stocking around his neck with an orange in each end. He got back about the time Hilman stopped the ring. When he pointed, this wiseacre stepped into the circle, and the chase was on.

Finally, in spite of his blindfold, Hilman captured his quarry who had been doing his damndest to sound feminine when he answered his pursuer's "Root." Then came the attempt at identification. Hilman felt his captive over gingerly, careful not to get too far down, or up—though he did manage to brush against the oranges a time or two. The kimono and the apron had him a little confused, looked like, but at long last he came out with, "Toots?" I guess we shouldn't have laughed like we did, because when his blind came off and he saw what he'd caught, he was terribly embarrassed. It could be, looking back, that it was the start of the two breaking up. Maybe not. Didn't matter much, anyhow, as it turned out. Seems that a while later down in the Melville store, Toots, either as a joke or because she decided to get the show on the road, accosted her swain with, "Hilman, will you marry me?"

There was a minute of silence as he thought it over. Then she damn sure got her answer, "Hmmm, God no."

Hilman got married later. But not to Toots. I liked her, she was a nice person.

Speaking of the Weslands reminds me. She was real nice to kids. Liked them, and we liked her. Anyhow when we first moved from the Billy Creek ranch down to Otter Creek and were living in a two-room, dirt-roof cabin, I had some sort of a falling out with Mother and Dad one evening. I forget why, so it sure couldn't have been a spanking matter. It came to where I finally announced that I was leaving—that I was going over to Weslands because she was nicer to me than I was being treated around home. To my surprise, Dad told me, "Go ahead." Then, to my chagrin, Mother fixed a sandwich, got my toothbrush, a pair of my drawers plus a shirt, put them in a bandana and tied the corners across the top. Looked like they didn't give a damn about me, so by the time it was handed to me along with my hat, I had worked up a good mad. Didn't help a bit either when I was told, "Say hello to the Weslands for us. Goodbye."

Well, it was night by then, and I had about six, seven miles to go, but I mumbled goodbye in as surly a tone as I dared and left. It was only a couple of hundred yards up to the lane, but it was dark as the inside of a black cat, and it

took me some time to make it. Then when I'd gotten about
a quarter-mile away from our gate, the damn coyotes
tuned up. I suppose there weren't more than a couple, but
they sounded like a whole pack. And over south where I
was headed, too. Hearing them, I got to remembering how
the wolves howled in the pen up at Billy Creek and the
stock I'd seen that were wolf-killed—and my powder got
damp. Right quick I made a change in my plans and
direction. Didn't take nearly as long to get back to the
house as it'd done to leave it, either. I was about played out
when I burst in the door and got it slammed shut just in
time to thwart the lead wolf, who I knew was at my very
heels.

Dad looked up from his paper, "Back already? You
couldn't have stayed long."

I was too out of wind to answer, but I decided then and
there that home was damn sure the best place to be. And I
haven't changed my opinion in the sixty-odd years since
then.

Teddy Knede was another well-known figure in this
area. He was a Swede, which in itself was uncommon in
these parts—Swedes and Norwegians not being what a man
would classify as boon companions. Matter of fact, a
popular story around here had to do with a violently insane
Norsk who escaped from an asylum in the old country and
just flat disappeared. He was in such bad shape that a
widespread hunt was made for him, but finally, after
combing the fjords and mountains for several seasons, the
search was abandoned. Then about five years later the guy
turned up in Sweden—as vice-chancellor of the University
of Uppsala, the absolute center of Swedish learning!

Perhaps the Norsks got used to him or something, but
Ted was around here for years. He was a stonemason, and I
imagine he must have built damn near every fireplace in
the country, as well as having a hand in most of the stone
buildings. He built them stout, for I know of a fireplace

back in the Crazies that still stands, though the cabin around it burned down nearly seventy years ago, as sound as the day he finished it. Another feature his fireplaces had in common was that they didn't draw too well. All of them smoked a little. Could be the smoke shelf he favored or some other particular design of his, but hell, if everybody has a fireplace that smokes some, you get used to it. But if there were only a couple that smoked, the people who had 'em would be unhappy, and those with the outfits that drew well would be smug. So it's probably better like it is. And has been for a long time.

Teddy was an old man when I knew him: big, powerful and, from years of handling rocks, hump-backed. It sure didn't hurt his ability to work. Many's the time I have seen him lean over, pick up a chunk of rock that would bother two men and swing it into place. To watch him study a rock, then hit it a time or two with his stone hammer and have it break clean along the line he'd chosen was something that always fascinated me. I asked him how he could do it, and his answer was, "Ay yust look t'rough de hide an' see vere he vants to break. Dat's all."

He was a great drinking man when he wasn't working and had a favorite ditty he used to sing when he was in his cups.

"Oh, Ay built de Crazy mountain'
An' Ay built dem dere to stay,
Long before Columbus
Awer anchor' in de bay."

In the open cirque on the side of Big Timber Peak, across the canyon above the dude ranch, the snow lies a long time. When most of it has melted, by about mid-August of an average year, the remnants stand out in the gullies as a vee and a reverse cee (ↄ). Our horse brand is VC so I jumped Ted out one evening when he was celebrating at the Cort Bar in Big Timber. I asked him why in hell, when he was building the Crazies, he hadn't gotten our brand right. "Ay got drunk on de yob vun day an' de bran' coom out assbackvards," was his answer. "Den ven I sober up she's too late."

He studied a minute and burst into song. The first two lines were his usual, but then:

"Ay got de Wan Cleve bran' on wrong
An' she von't go away."

Dad always favored horses that had some get up and go. His buggy teams, when I was little and we lived up on Billy Creek, sure qualified. They were exciting propositions, those buggy rides with Mother and Dad. Never just with Dad, for Mother had to be there to keep an eye on me. Dad damn sure needed all his attention for the horses. With good reason, for not only was the trip itself exciting—especially to start with—but getting ready for it was pretty complicated. First, the rig was backed into the barn behind the door. When the team was harnessed and hooked up, Dad would get up into the seat with the lines. Then Mother and finally I was handed up and ensconced between them. When everything was all set, a man would get hold of the door and when Dad nodded, would open it. The track was always greased so he could do it right, by God, now, too. It must have been just about, as I remember, like a saddle bronc quitting the chute, only that the team busted out the door at a hard run instead of bucking. So the first part of our jaunts was always a little rapid.

After they'd made a half-mile or so, things would quiet down, and from then on out Dad would have a good buggy team. A little warm, maybe, but traveling sons of guns. I sure liked those trips, especially their starts. The guy at the door, usually Riley Doore, would whoop as we went past, and I'd pipe up and answer him. It was fun!

An incident from one of those outings still sticks in my mind because it spooked me. In those days the average ranchhand wasn't long on refinement, particularly when he was at work. Naturally I tried to be like the men, and

Mother sure wasn't happy about some of the things I picked up. Nose blowing, for instance. I favored the men's method—lean over, put a thumb against one side of my nose and let 'er rip. Then swap thumbs and nose sides and blast again. Worked fine. Efficient, made a nice loud noise and, when I got the hang of it, not too messy. Mother, though, took a real dim view of the procedure. I got my britches warmed a time or two and was forced to pack a handkerchief, but I wouldn't use it unless it came to a case of root hog or die.

Well, this day after the team had their warmup, we spanked up the long flats past Battleship Butte and dropped into a draw at the head of the north fork of Otter Creek. There was a shack built into the slope toward the bottom, sort of half-dugout and half-cabin. A man I'd never seen before came out. Dad whoaed up, they started to talk, and my curiosity exploded. I'd never heard anybody that sounded the way the man did. Not with an accent—I was used to that—but more like he had a mouse, or something, up his nose. Sort of fuzzy sounding. Way later I learned that he had what they called catarrh, but even if I'd been privy to the fact then, I'd have been fascinated. I watched and listened a while, then turned to Mother, "Hey, Mom. . ." and she nearly broke a rib with her elbow. So I tried Dad. Mother shut me off with her elbow again. I subsided, but when Dad clucked to the team, and we pulled over the rise out of sight, I had another stab at it. "Hey, Mom, why did that man talk so funny?"

I got my answer this time. "Because he didn't blow his nose often enough when he was a little boy and wouldn't use a handkerchief when he did." I turned over a new leaf from then on out, for damn sure, when it came to nose blowing.

That incident from way back reminds me of one morning at the dude ranch a few years ago. We wranglers had come up from the corrals where we'd been graining and saddling horses and were eating breakfast, as usual, with the cabin crew. All of a sudden one of the latter, a girl from back East, asked plaintively, "Where are the napkins this morning?"

I had a boy working for me that year who was a dandy youngster, but he'd been raised pretty much at the head of the crick, and some of his rough places hadn't been smoothed down much. So, serious as a tree full of owls, with his expression reflecting complete bafflement that anyone would fuss about the lack of a damn paper napkin, he looked up from where he was shoveling in hotcakes and eggs. "You got a shirt sleeve, ain't you?"

I had a hell of a time keeping my face straight, and the girl was plumb speechless.

Another deal that involved a buggy occurs to me. Shortly after we moved down on Otter Creek, the school board decided to bring the schoolhouse over from the Gunderson corner to Dry Creek. This was right near home, so my sister and I went to school afoot instead of ahorseback. It also meant that the Anderson kids, who lived a couple of miles below us over on Rye Creek, came right through our place every school day. There were five of them, so they drove. When there was snow, they'd use a little sleigh and if the weather was real rank, they'd leave their driving horse in our barn during the day, for there wasn't any sort of shelter at the new school location. They used a buggy when the ground was bare, naturally. They had the latter one fall morning when Dad was threshing a patch of oats he had out just north of the house. The rig was humming along, Hilman didn't blow the whistle or anything like that, and Anderson's old horse—who'd seen threshing outfits many's the time—looked over at it, snorted and dropped dead. Just like that, right between the shafts.

It was so fast that it took a little time to register. Then everybody cried, from the oldest down to the littlest.

Naturally, for every one of the Andersons had known and loved their old horse all their lives, for I am sure the horse was a good deal older than any of them were. To tell the truth, my sister and I weren't in a hell of a lot better shape, so it was a weepy arrival at school. It was catching, for at intervals during the day one or the other of the Anderson girls would begin to sob and pretty quick every girl in the room would be sniffling. The boys tried to be manful about it, as men should, but it was a sad, sad day—the worst I can remember in my eight years at school there. Our friends had lost a dear friend, so we all grieved with them. Besides, we all had horses, too.

Today I suppose a shrink would say that it was a traumatic experience which would leave a lasting mark. Sure, it was tough, but ranch kids are raised close to death, and new life, and every one of those children today is a rancher or a rancher's wife. Good ones. Ranch kids learn early.

My uncle Jack Scarlett had the first car in this part of the country. I was too little to remember it in action, but I've heard lots of stories about it. It was a red Franklin chain drive. You got into it by way of a sort of tailgate door, though I don't know if the back had a seat like we have now or whether there was a seat along each side. In light of the things that happened, I'd say probably the latter. Of course, it didn't have a top. Jack, a racehorse man and polo player, drove it like he was on a good horse, like the mill tails of hell. Especially after Aunt Helen died, and he got to drinking pretty heavy. One afternoon he, Charley Dugro (another uncle), John (the ranch cook) and my cousin Allie Langston made a trip down to Melville in the car. I be damned if I know how it came about that a little girl was along with that caliber trio, but she was. Probably, Jack and Charlie were just being nice and giving their little niece a car ride. They were good guys, though not what would be considered real responsible.

When they got to Melville, they finished whatever
business they'd come for, bought some candy for Allie, put
her in the back of the Franklin with orders to stay there
and dropped into the saloon for a short one. Happened
they met a couple of friends in the establishment, others
kept dropping in and so the short one got longer and longer.
It happens. One for the friends, then one with each of those
friends—you can't insult a man by not drinking with him.
Then, in Melville, the house would buy a round. Not
celebrating, really. Just a friendly drink with friends and
an earnest, intelligent discussion of subjects of immense
importance. Then, hell, one for the road—but a bird can't
fly on one wing. Then suddenly, the stark realization that
you were due somewhere else quite a while ago.

Jack finally remembered that he had the cook with him;
the cook who was supposed to be cooking supper. So he and
Charley got John into the car, and they headed for the
ranch. Jack was in a hurry, for he knew that Granny, his
mother-in-law, would be on the fight about her cook. If
he'd been sober, he'd have known that John was in no
shape to cook supper, probably not even breakfast, but he
wasn't. So he cut across country to save time with the
Franklin wide open.

They'd made it about halfway when they noticed Allie
was talking steadily, and finally it dawned that she was
trying to tell them something. Jack slowed down enough so
they could hear what she was saying, and what they heard
was, "John's gone, John's gone." Repeated over and over in
a happy, little girl voice. They craned around to the back,
and, by God, she was telling the truth. John *was* gone.

Allie couldn't elaborate, so they turned around and
started back the way they'd come. Sure enough, after a
mile or so, there was John sleeping peacefully beside an
irrigation ditch Jack had crossed at top speed. They poured
him into the back and headed for the ranch again. More
circumspectly, I imagine, for everybody was still aboard

when they arrived. I never did hear, though, what Granny's reaction was, worse luck. For she could be a little raspy about some things, as we kids found out a number of times.

That wasn't the only time Jack lost a passenger. He was coming back from Hunter's Hot Springs to Big Timber with a load of people and shed my uncle Tom Blakeman. Once again they had drink taken. Which was probably real lucky for Tom, for otherwise he might have gotten stove up. Doesn't speak too damn well for the roads of those days, either, for he lost Tom on a main highway.

Another time, Jack was parked out in front of the Grand Hotel in Big Timber when a herder happened by. Evidently, it was the first car he'd ever seen, for he walked around it several times at a respectful distance, muttering to himself in Norwegian. Jack, pleased with his interest, took him in tow and showed him all about it. They looked the engine over, examined the running gear, fiddled with the steering wheel and peered into the tonneau (that was what they called the back seat in those days). Finally, Jack said, "Would you like to take a ride with me?"

"Oh, nay, nay," was the hasty answer. Silence. Then, "Does she awer tip ower?"

"No, not yet," Jack answered, "but I suppose it could."

The Norsk thought a minute, then observed seriously, "Vell, ven she dus, Ay bat dere vill be vun hal of a crash."

At least three generations of our family have not been mechanically minded. Gramp tackled driving a car once, but never again. Dad could handle pretty nearly any sort of machinery—provided he found out how to start it—but if it broke down he had no idea how to fix it. Me, too. If something mechanical quits me, and they seem to have a propensity for it, all I can do is look in the gas tank. If that's in good shape, then I suppress my urge to work the damn, sneering headlights over with a rock and go call a mechanic. All three of us, though, could handle and do a fair sagebrush job of repairing any piece of machinery that had an evener, bull wheel or pitman stick. We savvied horses and what they were hooked to. My son, Tack, is different. He can fix any damn thing from a vacuum cleaner to a Caterpillar. It's real handy and saves money. But I'd rather be me. Life is much simpler.

Oh, Dad had cars when I was a kid, all right, and he normally drove them with the same élan as he did his buggy teams. Tosten Stenberg, who worked for the outfit for years, was quite a jack of all trades, liked engines and such and did a popping good job of keeping them running. Dad gave him plenty of practice. Trouble was, in order to make things go, you had to know just which of "Toosty's" homegrown gadgets to fiddle with. He was a tinkerer from away back.

We made a memorable (at least in my youthful estimation) car trip with Dad one time. We still lived up on Billy Creek, and I was pretty little, but it's as clear in my mind as though it had happened yesterday instead of nearly sixty years ago.

I haven't any idea what make car we had, but it had a top you could put up. It was up this day, glory be, or things would sure have been worse. We'd gone to Melville and were on our way home up the Settlement lane. Flat tires, inevitably, were a sizable part of car trips in those days and on those roads. Dad was pretty cavalier about flats this day. Seems he'd gotten his tubes filled with some sort of liquid rubber that was supposed to seal punctures the minute they occurred. At least, that's what the advertisement had said. Anyhow, since we'd had no trouble so far, he laid it to the stuff. "Hell, we've probably already had some punctures, for we've never gone this far without one before. It must work to beat the band."

Well, maybe it had, but if so, something got out of kilter in a hurry. Just below the Lutheran church a string of what looked like molasses looped up on Dad's side and down over the windshield and hood. By the time he got the outfit stopped, his side of the car looked like it had a fishnet strung over it. It smelled, too—sort of like the goo that you squeezed out of that tube that always came in the puncture repair kits everybody carried.

Pop opened his door and promptly picked up the threads. They wouldn't brush off him, and when he tried to pull it loose, his fingers got stuck. He muttered something that made Mother give him a hard look and told us brightly, "Must be a pretty bad leak, but it'll seal in a little."

He waited ten minutes or so, and when we started up the lane again, everything seemed fine. But when he picked up speed, the threads began showing up, so we stopped and waited once more. Longer. When we started, it was a repeat of the first try. I heard what Dad said this time: "Sonofabitch!" And he tramped down on the gas.

The further and faster we went, for Pop was on the fight and was out to *make* the stuff work, the bigger and juicier the strings got. Hell, they looked like clothesline. Finally, it got to where he couldn't see out of his side of the windshield. He wasn't about to poke his head out so's he could see, either. I doubt he could have, anyhow, so we stopped, got out the yon side and stood looking at the mess. Slowly, the red in Dad's face paled, and all of a sudden he began to laugh. Then Mother, though not nearly as wholeheartedly. When I was sure it was safe, so did I for I'd been busting to for the last mile.

I don't know exactly how the car got back to Billy Creek, for Mother, very wisely, wouldn't let Dad mess with it. So he walked over to Hart's, called the ranch and a while later Toosty showed up with a team and rig. We went home in it, and Toosty stayed with the car. The next time I saw it, several days later, it was cleaned up. All except the smell, that is. That lasted for months, in spite of all Tosten could do. Finally it wore off, or maybe we just got used to it.

I never heard Dad mention his hotshot tube filler either,

except years later. Pretty sheepishly, too. My bet is that Toosty was told to get rid of the works; the stuff, tubes and all. At least air, when it escaped, didn't decorate things, and a man could put it back in with a pump, to boot. Still, it was funny as hell, and I still chuckle when I visualize it.

Dad as a young man

Dad

Dad roping calves for corral branding

Dad on Keogh

Threshing

Dad (left) and Spike

Dad and Shelly after moving cattle in Section 16

V reverse C on face of Big Timber Peak

Team and buggy crossing the creek

Spike on Panama, 1918.
Spike was 6 years old.

Panama, Aunt Allie and Spike
(in his wooly chaps)

The 1905 Franklin with Jack Scarlett driving

Johnny Christenson of Two Dot fanning

Homemade sled, Spike and Bullet and Babe

Keeping bridle bits warm in winter

Ground blizzard drift at the Butte Ranch

Feeding

Come and get it!

Looks like a long winter

Bunkhouse

Those PY Whites

Hi Whitlock riding Christian Pride, 1951

Shelly hunting bulls in late November with Barbie

Melville's main street in early 1900s

In Melville

Horse racing in Melville's main street—late 1890s

Sunday race

Two real hands

═══6═══
UNDER THE CRAZIES

MY GRANDAD had a story from the early days about a man who was going to be hung. Everything had been plumb legal; he'd had a trial, was found guilty, sentenced and the date of the execution set. Back then, a formal hanging was something of a community social event. So when the big day arrived, the square in front of the courthouse and gallows was loaded with spectators, some of whom, especially the families with kids, had brought lunches and were picnicking as they waited.

There wasn't all this appeal nonsense in those days, so about the time everybody had finished eating his sandwich and before the patrons of the local saloons had had time to get too drunk, the central figure in the proceedings appeared. Under guard, he was escorted up the steps and onto the trap where he was readied for his long trip. Before he was blindfolded, though, he was asked if he had anything to say.

Now this old boy hadn't ever seen as many folks bunched up at one lick before, and he must have gotten sort of carried away with all the excitement. Anyhow, his answer was, "To tell you the truth, if it wasn't for the honor of the thing, I'd jest as soon not be here."

Now I don't know if the story is true, but it sure typifies the slaunchwise sense of humor so ingrained in this region and its people. It is still very much alive, too. Take this incident, which happened a couple of years ago, for instance.

A local man, who had ranched and batched "at the head of the crick" for years finally sold out, bought a little outfit down near the Yellowstone and semiretired. In the course of time, he married a widow woman. She was a nice person, though pretty well up towards the lead when bossiness was given out, but the two were as happy as bugs in a rug. A couple of years ago, before gas got so high, they took quite an automobile trip one fall. After they got home, a neighbor who happened to drop in asked them all about the trip. Where they'd been, what they'd seen, had they enjoyed themselves and so on. They were real enthusiastic about their travels and the scope of country they'd cover-ed. The neighbor finally remarked, "Sounds like you went a lot of miles. Who did the driving?"

There was a moment of silence, then the husband answered, "Well, I held the wheel."

I remember another classic understatement. We were trailing to Big Timber to ship one fall. On the way, we had to pass a homesteader's outfit. Not a very prosperous layout either. Just a little barn, house and chicken house with a set of clotheslines strung between the latter two. The missus had evidently done her washing that morning, for the lines were loaded.

They had a dog. Now, while I don't mind a dog that watches out for the house, damn if I like one that goes out of his way to cause trouble. This one did. As the tail end of the cattle strung by, he slipped out without a sound and heeled one of the saddle horses, hard.

This pony had been a little broncy that morning but by this time had gotten bored with cow critters and was dozing as he ambled along. When the dog nailed him, he woke up right then and got *damn* broncy. His rider was a good hand but had been all gapped open, too, and got in a storm in a hurry. He was down on the withers trying to get his pants in the saddle and gather his slack, when he looked up and saw that clothesline coming. I've never seen a man swap druthers faster. He quit worrying about his slack and went to getting up to where he wouldn't catch the rigging under his chin. Talk about "tall in the saddle"—it looked like he was standing on his toes in the stirrups! Back straight, chest thrown out, jaw like a rock, he hit those lines at the top of a jump.

The first two he took across the chest; the final two, when the pony hit ground, hooked on the saddle horn. All four broke in a series of high-pitched twangs, and away they went in a cloud of clothes. All you could see were two heads, the horse's and the rider's. All the rest was a billow of things going to hell in a handbasket in a hurry.

The bronc must have looked over his shoulder about then and figured something had sure gotten him, for he quit bucking and tried to outrun whatever the hell it was. Getting things under control was anything but easy. To begin with, we were laughing so hard we couldn't operate plumb efficiently. Then, too, our horses didn't like what he was carrying a damn bit better than the bronc did. In time though, the clothes pretty well wore off the lines, and the rider got the two off his chest and back over his head. By then, his horse was getting tired, so we got him stopped.

Part of us went around the cattle who, naturally, had had a nice run. A couple of others coiled the lines on their saddle forks, and then all of us went to gathering clothes. All the way back to the house.

Our reception there was mixed. The husband had seen it from the barn and thought it was a hell of a joke. Not so his wife! Dad apologized, but she ignored him and proceeded to cuss the living daylights out of the unlucky rider. And she could damn sure cuss, too. Couldn't blame her. She'd had to pump and carry all that wash water, scrub the clothes on a board—and everything was dirty again. She hadn't seen the dog, either. Just figured here was a cowboy showing off. She even took a run at the poor guy with a broom, which resulted in another bronc ride. When things cooled down some, he looked at her earnestly and got in his first, and final, word. "Ma'am, I sure didn't plan it."

That old pony hated dogs from then on out. Laid for them until he could get a good bust at them, and it was a couple of years before his rider could pull out a handkerchief without having a runaway on his hands.

Funny thing but the Melville humor seems to rub off on outsiders sometimes, if they are around here long enough. To be a little screwball to begin with helps, too. A crying example is a friend of mine from the East Coast who first came to the dude ranch as a youngster. He has been back many times since—on college vacations, with his bride, with his family and, just lately, with his first grandchild. Several winters ago, I got a note in the mail: "Once again I have decided to undergo your inadequate food, exorbitant prices and filthy accommodations. For August." It was unsigned, but with it was a $1,000 bill. Just like that!

My reply was, "The bank has pronounced it genuine, so you have a week's rent paid," and I didn't sign, either.

About mid-July, another note arrived: "Warn the local liquor purveyor you are expecting a dry friend."

My answer was, "I already have," and on the first of August he and his family arrived at the ranch. He'll do!

I'm against dudes on our outfit using spurs. We've been running horses for right at a hundred years, have raised all the stock in our dude string, and they are good ones. Hell, spurs aren't needed with them. I use 'em, sure, but then I'd feel naked if I didn't wear spurs ahorseback—though I've seen times when I damn sure left them off. Besides, I know how to use spurs. So when a guest of long-standing wrote for reservations and added casually, "I'm bringing a pair of spurs in hopes of occasionally achieving a trot," I made a mental note to square up with the smart booger, friend or no.

He brought the spurs, sure enough. He hadn't figured on my insisting he wear them, which I did. To top it off, I mounted him on a big, good, touchy gelding, and after "achieving" a good deal more than a trot, he left the spurs up in his cabin for the rest of his stay. He was all right, damn him.

I guess the remark that will stick with me the longest was made by a woman from the horse country in Delaware. She and her family had been at the ranch for a number of years. A dandy outfit and good friends. Happened she came down to the corrals early the morning after they'd arrived. I was tickled to see her, we talked a little, and then she asked, "What'll I ride this year, Spike?"

"How about Dawn?" I suggested. "She's a hell of a lady's mare."

"Fine!" was the enthusiastic reply. Then, "What have you got for Bucky (her husband)?" and without batting an eye said, "Old Nutcracker again?"

I guess my face must have shown something, for she hastily added, "Well, that's what *he* calls him."

Years ago, it seemed like the Melville country always had somebody visiting at one of the ranches. Easterners, mostly. Friends, friends of friends, children of friends. Boys or young men who came out to spend some time "working" on a ranch. A delicate way of putting it! Others were inflicted on the ranchers to, hopefully, "learn responsibility," or, in a few instances, I am sure, to do their drinking a long ways from back home. The Donald outfit had a few, so did Hart's Dot S Dot, and we drew our share. However, Dad put up with damn little hanky-panky, so our rollicky ones didn't seem to stay very long.

Most were pretty likeable, by and large, and fit right well into the local idea of joie de vivre. One in particular, at Hart's where he tried Mr. Harry's patience to its utmost limits, was outstanding, even in Melville, for his failing for booze. Doc was his name. A good-natured cuss, intelligent, but interested solely, it looked like, in the next drink. Why he didn't get killed, the fool things he did, I'll never know—perhaps because, as Gramp always said, "God takes special care of children, locos and drunks." He sure needed special care, or something.

Like the evening after a Melville rodeo when he, either surfeited with moonshine or maybe too plastered to be able to swallow, decided to call it a day. So he bedded down right in the middle of the town's one and only, and therefore main, street, spang in front of the busy dance hall. Carefully he moved a rock or two, lay down in about three inches of dust, turned around a time or two like a dog getting settled, pillowed his head on his arms and went to sleep.

It was a sight! Cars would come boiling down the street, spot him and swerve in a cloud of dust. There were a few real juicy near misses, but he was as oblivious to it all as a babe in his mother's arms. Didn't even stir when "Stub" Rudd, as usual, had a falling out with somebody, and the two fought around and over him. Doc had lost his hat, looked like, before he bedded down. At least, he didn't have it with him, so in short order his face, hair and everything else carried a heavy layer of dust. Matter of fact, he resembled an old wadded-up tarp that someone had lost out of their outfit. Either that or a pretty bad bump in the street, which may have been part of the reason he wasn't run over. Also the fact that Melville always respected the individual's rights—so long as they didn't run contrary to other people's rights. Every man should kill his own snakes. So if Doc wanted to take a chance on sleeping in the middle of the street, it was his choice, and right, to do it. If he got crippled, well, hell, nobody'd put him there. A drunk is damn hard to hurt badly, anyhow.

He must have made it through the night's festivities safely, for he was still there (a source of great interest to the town dogs) the next morning. In any event, when he woke up and pulled himself erect, he showed no scars. Dirty as hell, yes. Dust—and mud where the dogs had favored him—but no noticeable contusions, and time he'd had a snort or two at McQuillan's, he was back to normal.

There was one time, though, when Doc got pretty badly jarred. Not physically, but mentally, or maybe it could be called psychologically. Hart had mounted him on a nice roan mare they had. She was gentle, in foal, and by the time he'd ridden her a month or so, it got damn hard to keep her from heading towards Melville whenever he got aboard. I've heard it told that if Doc got on her any time of day, up to along towards evening, she'd automatically start for the saloon. After dark, however, she'd head for the ranch. Particularly if her rider was stepping high. She was a good mare. Stood nicely for him and traveled with an ear and an eye back so's she could step over under him if he started to topple.

They came back to Hart's that way one nice spring evening. Doc tied her by the back gate while he went into the house for something or other. When he was done, he went out to take care of her.

In a minute he was back, his face drawn, and his eyes scared. "Uncle Harry," he asked hesitantly, "would you please come outside with me?"

Mr. Harry followed him out to the back gate, and there, big as life, was a fine newborn colt sucking the roan mare. "Uncle Harry," implored Doc, "I really *see* a colt, don't I?"

Mr. Harry assured him that he did indeed; that the mare must have foaled while he had been in the house, and Doc finally let his full weight down. I heard that he stayed off the booze for the next three or four days, though, and he got a new saddle horse.

We had a visitor from somewhere back East one winter. I don't know where, but I'd bet my bottom dollar it was New York and he'd probably gone to Yale. Funny duck; serious, pompous and ponderous to the point that he'd have made a horned owl look plumb giddy. Looking back, I'd say he must have had an idea that Westerners, particularly ranchers, were a pretty uneducated, ignorant lot. At least, whenever anyone expressed an opinion, no matter how innocuous, he'd puff up like a sage hen on a stomping ground and clear his throat importantly. Then, in a resonant tone, he would pronounce, "There are *two* schools of thought concerning that," and launch into a detailed discussion of both of them. Reminded me for all the world of an old-timey preacher discoursing on hellfire and brimstone. Most of the outfit, when we got used to him, enjoyed needling him with locoed ideas and then slipping off to leave him expounding to himself. Dad took a somewhat different view of things, and I noticed, much to my glee, that no matter how many schools of thought were propounded, Dad's school seemed to prevail. If there was work to be done, it did it damn quick, too.

We had another young fellow spend a winter on the ranch who was a horse of a different color. His name was Harry, and he was a peach of a boy. We all liked him, and he fit in well, but took the damndest, most literal view of things of anyone I have ever seen. A crying example of this habit came to the fore when Mother baked a cake for the Halloween party at the Settlement School.

She had done a fine job. It had orange icing, trimmings from hell to breakfast in keeping with the spirit of the occasion, and we were all admiring it when she suddenly said, "Wait a minute."

Then she rummaged around in the cupboard and dug out a lazy Susan deal, put the cake on it, wound the key on the pedestal and the outfit went round and around while its innards gave forth with "Happy Birthday."

"I think that would be nice to use at the party. Don't you?" she asked proudly.

Harry watched intently as the cake revolved to the strains from the music box. Then he shook his head. "No, Mrs. Van Cleve," he answered seriously, "I think it would only confuse the children."

The Donald outfit down on the Sweet Grass also contributed to the local pool of imports. I don't know whether he was a regular hand or was just "working" there. I sort of imagine it was the latter, for I can't imagine an outfit paying him wages, especially of the size it took to keep him operating, for he cut a wide, though meandering, swath. His name was McClure, but he went by "Mcoola." He wore a black, go-to-hell hat, had a glass eye and an overwhelming thirst. Consequently, he put in a lot of time ahorseback to and from the Melville saloon.

The "from" was the trouble, for by then he'd be pretty limber in the saddle. Mostly, when he fell off, it was like a dishrag, but this time, just above Green's gate to the Settlement lane, he must have lit hard enough to pop his eye out. He was sleeping peacefully when a neighbor woman came up the road. She was not native born, and though she'd been in the community for years, she'd never gotten the hang of it. But like a good Samaritan, she stopped her car to keep from running over the body and got out to investigate. To her relief, she found that it wasn't dead, but to her horror she saw that it was missing an eye. As it was the nearest place, she drove into Green's with the tale of her find and asked them to call a doctor. They asked if the man had been wearing a black hat, and she said yes. So they called Donald's while she hurried back to give what succor she could.

Finally a couple of Donald's men arrived, along with McClure's horse, which they'd found standing at their gate. The lady met them, wringing her hands about the horrible eye accident. "Hell, Ma'am," one of them assured her as they boosted the limp form into the saddle. "It's only 'ol Mcoola, an' he's got a matchbox full o' eyes in th' bunkhouse."

I don't believe she ever got plumb over the deal. Acted like it was all McClure's fault, which, come to think about it, it was.

She got herself in a jackpot another time. With the ladies of the "Settlement." Seems she gave a big coffee party for them, and, as the saying goes, really put on the dog. She could do it, too, because before she married a local rancher, I understand she'd been dean of women at a sizeable Eastern college. So the party was a classy one, and the ladies were quite impressed. Along toward the end of the festivities, when everyone was full of coffee and cake and the talk had turned, as usual, to local gossip, the name of a woman not present came up. A discussion of her went on, but ended abruptly when the hostess leaned forward and confided, "Well, I think she is a s-n-o-b."

Her guests, mainly nice Norwegian women and not especially sophisticated, didn't savvy just what she'd meant. It damn sure sounded like something else, though, and the party broke up pretty fast. Took a long time before she got herself squared up with the ladies, too.

Years ago, I took a trip around the Crazies with our forest ranger—back in the days when the breed deserved the name "ranger." We were gone a couple of weeks, for we prospected every canyon to its head, mended trail, sized up the feed and the livestock on it and camped wherever night happened to catch us. Towards the end of the trip, we pulled out of the Smith Commissary on the head of the Shields River one afternoon; we'd stopped there to chirk up our grub supply. We were headed north, so we took the old road between Target Rock and Virginia Peak to Forest Lake. It was late and full dark when we hit the lake. There seemed to be some sort of an outfit down the shore a ways, but we hunted up a level spot, picketed our horses, had a bait of crackers, sardines and canned tomatoes (thanks to our stop at the commissary), rolled out our beds and turned in.

We woke up the next morning to the damndest racket. Sounded like school'd let out, and it all seemed to come from a camp about a hundred yards below us. Our horses were curious as all hell about the commotion, and so were we, so when we'd eaten, we strolled over to see what was going on. There seemed to be kids everywhere, so at first we figured it was some sort of a children's camp. It sure could have qualified, but it wasn't. It was a rancher and his family from down on the Musselshell. The missus was sweating over a big old kitchen range set up under the trees while her husband was stretched comfortably in a hammock slung between two of them. The rest were kids, their kids, I guess, and they ranged all the way down from about my age to the baby who was gurgling happily as it stewed in its own juice in a basket beside the stove. I couldn't get a standing count on them, but there were a lot. Whooping, crying, running this way and that, or standing, in the case of the littler ones, staring at us and sucking their thumbs.

The man waved from his resting place: "Howdy." Then to his wife: "Git some vittles." We told him we'd eaten, so he sank back, and we chatted. Sort of tickled me, for every time a youngster happened to get close to his hammock, the man would bat at him or her. Reminded me of a bear. Finally, the noise got the best of him, for he filled up his paunch and let out a bellow: "Hesh up, dammit!" Then when the tumult waned: "You thank you're agoin' to th' mountings next year. Well, you ain't!"

I've never forgotten the incident, nor the words, and over the years it has become a favorite saying among the family and a few close friends. When Barbie was in college in Omaha, she and a classmate on their way down one fall detoured through the Badlands. A few days later I got a postcard from her. On one side was a picture of a wicked-looking pinnacle entitled "Vampire Peak, S.D. Badlands." On the other, in her handwriting, "You thank you're agoin' to th' mountings. Well, you ain't!" After all, she was raised under the Crazies.

When Bob Langston was wrangling at the dude ranch, he and I took some guests up to Crazy Lake at the head of the South Fork of Big Timber Creek for some fishing. Grosfield's camp tenders had blazed a trail as far as the upper meadows. On up the creek and then to the top of the glacial step where the lake lies, there wasn't what could be called a trail. Ken Pickens and I had camped at the foot of the step a few years earlier and had scratched out a track up the face, so we could get a packhorseload or two of trout fry planted in the lake. But it was still pretty rollicky to get there.

Anyhow, about halfway from the meadows to the foot of the hill, we caught up with a rancher and his family from way down the creek. Haying was over, and they were going camping for a few days. It was a baling wire outfit, if I ever saw one. Afoot, without even a packhorse. Each kid, even the littlest one, had a load of camp equipment, spuds, canned goods or something. Pretty damn big loads, too. But the man and his wife had the clincher—a couple of bedrolls and a tarp slung on a pole between them. They were having plenty of trouble with it, even though they were still on comparatively level ground, and I winced to think of what they would run into above.

We whoaed up to say howdy. They were all dripping sweat, and the man and his wife were pretty ginger when they took the pole off their shoulders and set down their load. Even though the ends of the pole were padded with gunny sacks, they were getting a little sore. It was such a pitiful looking bunch—I knew what was ahead, to boot— that I was almost ashamed of sitting up there on my horse. They not only eyed us enviously, but damn near accusingly, so we were all relieved when we went on and left them behind. Which was right quick, for I sure didn't want those shoulders to get cold, or they might be like a couple of work horses. Cold shoulders and work won't mix.

We got to the lake, fished for a couple of hours, then tightened our cinches and started home. When we got to the top of the step, I stopped, and everybody got down. The damn hill was bad enough to come up; going down is always ranker. So I figured we'd better lead our ponies until we got to the bottom of the canyon. Except at the creek crossing. We'd ride to ford that.

I was just starting over the edge when I caught a movement below us on the lower flank of Crazy Peak on the yon side of the creek. It was the footbackers, and I, savvied right away what had happened.

The ford was easy to miss. It was on a rock ledge at the head of the falls, and a chunk of the cliff that had come down from the mountain sort of hid it. It'd fooled me when I worked out the trail. The going looked easier along beside the creek, so I'd taken it for a quarter-mile or so until I'd run up against the peak. There the slope turned ungodly steep. Too steep for even slide rock to get any footing. Just solid rock or slick grass. Sure no place for a horse, so Ken and I'd backtracked and finally found the spot where we could ford. God knows the terrain we struck when we got across was rough, but it beat what we'd left all to hell, and we puzzled out a way to the top that horses could handle.

This outfit must have missed our tracks to the ledge and had been suckered, like I'd been, up along the creek. When they'd hit the steep slope, they'd tackled it. Probably because they thought they could handle it afoot, or very possibly, they didn't cotton to the idea of backtracking and losing altitude.

They weren't doing too well. As we watched (it was easy to follow them on the raw, open slope), one youngster slipped, lost whatever he was packing, and it rolled down into the creek and disappeared. It was too far to hear, but from the man's gesticulations he wasn't too happy about the loss. Then, looked like, he and his wife got into an argument. Then another kid fell down but didn't lose anything. Except maybe some hide. Then Pa went down. We were able to hear the results of that, but not the words. Probably lucky we couldn't, for we had women with us.

Then Ma sat down. I couldn't tell whether it was a slip or
that she'd just decided, by God, she wasn't about to go any
further. So there! But it was getting on, and if we wanted to
make the ranch by suppertime, we had to go. "Poor devils,"
I said, and I meant it.

Then that damn Langston, with the detached, sanctimonious expression which usually precedes his smart
cracks, remarked, "You thank you're agoin' to th' mountings next year. Well, I 'magine you ain't."

I might as well finish with another of Gramp's stories. He
used to tell of a rancher friend he had years ago over on
Lebo. The guy hadn't had much schooling as a kid, but he
read everything he could get his hands on and prided
himself on the vocabulary he'd gotten together. He was
always using big words, but a lot of the time he used them
wrong. Got messed up with the way they sounded. Also, he
was a bachelor and courted every single woman in the
country. Gramp met him on American Fork one day. The
old fellow was all slicked up, driving a fine pair of roadsters,
harness all cleaned and buggy shining.

They both whoaed up, and Gramp, curious about the
turnout, asked, "Where you going, George?"

"Ain't goin'. Been," was the answer, in anything but a
cheerful tone.

"Sure enough. Where?"

"Oh, courtin' the school ma'am at Melville."

"How you coming along?"

"Bad. Hell, took her for a buggy ride today an' exposed
to her twice. Ejected me both times. Guess I better give it
up."

=7=
TWO DOT

I WISH I'd known Two Dot Wilson, for in a region peopled with salty individualists, he was among the saltiest. He was born in New York state in 1830, then spent some time in Wisconsin as a young man. In 1864 he came to the gold camp at Virginia City with a wagon train that had quite a fight with the Sioux on the way. He cut firewood for the camp that winter and in a stagecoach holdup the spring of '65 was robbed of all his money. Quite a bit, too. Two Dot told Gramp he recognized the road agent but kept his mouth shut at the time—which was smart. "But," he ended the story, "by God, I had th' pleasure of helpin' hang th' sonofabitch a little later on." So it could be that he was one of the vigilantes that cleaned up the camps at Virginia City and Alder Gulch. I don't know.

He started in the cow business along about '69. At the same time he came up with the brand that gave him his nickname. Two dots (··) on both hips for cattle and the same (:) up and down on the shoulder for horses. I understand that the first time he ever used the brand, it was put on with a kingbolt from a wagon.

In '77 he settled on the head of the Musselshell. Gramp and Dad knew him well, and I've heard stories about him all my life. He was big, wore a beard, hired good hands and he didn't give a damn how tough they were. He was tougher. He also was a first-class swearing man, tighter than Billy be damned and pretty sloppy personally most of

the time. He was so bad that they tell of one time back in
Chicago with a trainload of cattle a policeman picked him
up for vagrancy. The cop couldn't be blamed, for Two Dot
had on the clothes he'd worn on the trip and was wearing
one of those cheap cloth caps advertising some flour or
other—about like the caps so in favor with the team ropers
of today—for somehow he'd lost his hat on the trip.

Anyhow, on their way to the jail they happened to pass
the bank where Two Dot did his business, and he asked the
policeman if they could stop in a minute. After some
hesitation on the law's part, for he couldn't figure what a
damn bum would do in there, they stepped in together.
Immediately, the bank personnel scurried to inform the
president of the outfit. Naturally, for here was one of their
sizeable depositors. The president arrived and greeted his
client warmly with "and what can I do for you, Mr.
Wilson?"

By this time the cop was swallowing a little hard and
wondering just what sort of a box canyon he'd gotten
himself into. It didn't help any when his "vagrant" answered, "Just how th' hell much money I got in here,
anyhow?"

The banker batted his eyes, "I can't give you the exact
figure offhand, Mr. Wilson. Why do you ask?"

Two Dot drew himself up in all his unsartorial splendor.
"I just wanted to know if I had enough to pay this blue coat
sonofabitch off" and turned back to his erstwhile escort.
The latter was headed for the door at a high lope.

The old fellow used to complain about the hotel where
he and his cowpunchers stayed in Chicago when he shipped
in the early days. Seemed he'd nearly lost a man on
account of the gas lights. Those country boys from the
west would *blow* out the light when they turned in. It sure
must have taken some doing, too. I guess, though, a few
men died of it, and the hotel finally had to post a notice by
each gas light in the rooms warning the occupants not to

blow out the light but to turn it off with the petcock.

Incidentally, the word *cowpuncher* evolved when cow-men started shipping cattle by rail. Every man with a shipment kept a prod pole with him whether he bunked in the caboose or a drover's car. At each stop they'd go along the train and peer through the slats of the stock cars to see whether any cattle had gotten down. If any had, they were prodded, "punched," to their feet again to keep them from being trampled to death.

In '76, before he moved to the Musselshell country, Two Dot went back to Wisconsin and married a girl from New York state. Their courtship couldn't be called long or involved. Two Dot proposed, but she asked for some time to think it over. "Sure," was his answer. "Take your time. Th' train leaves tomorrow mornin' at ten. You either go with me or you don't."

She went with him. They were married early in the morning and took the train. She had no real idea of what business he was in, for somewhere in the Middle West, as she watched the farms with their milk cows go by, she asked wistfully, "George, do you suppose we can have a cow?"

He agreed, "Wul, Berry, I reckon we can rustle you up a cow." He sure should have been able to!

Mrs. Wilson's name was Harriet, but for reasons known only to him, he always called her "Berry." In '78 there were only three white women in the upper Musselshell valley, and she was one of them. She settled into ranch life as though she had been born to it. She also, small and quiet though she was, got her husband reasonably well halter-broken as to clothes and personal appearance but putting a stop to his cussing was a lost cause—though she never gave up trying.

Two Dot remarked to Gramp, apropos his wife, "A man get used to Berry an' she's plumb all right. House looks better, place looks better an' by God, I look better since we been married."

It was a lot more than that. The two were very much in love. It happened that she got terribly sick and there was no doctor closer than White Sulphur Springs. It was mid-winter, one of those old-fashioned Montana winters, but her husband was so worried that he wrapped her in a buffalo robe along with several hot bricks and headed for White Sulphur in a sleigh. Every so often he'd stop, open the robe and see how she was doing. Each time she would smile weakly and reassure him.

They'd gotten pretty well up along the Castles when he made another stop, but this time she just lay there, her eyes shut. He tried to waken her. No luck, so he started off again. He came into Checkerboard at a dead run, whipping over and under. They heard him coming, caught his team as he pulled up—though I doubt the horses were in any shape to run off—and he hurried inside. The lady of the house asked, "What in the world is the trouble, Two Dot?"

Tears running down his face, he blurted, "It's Berry. She's real sick an' I was takin' her to White Sulphur. But she's dead, goddam her, she's dead!"

She wasn't, though. They brought her in, stoked up the fire and got some hot soup into her. Then with the bricks reheated in the oven she was wrapped up again. Two Dot borrowed new horses and made it to the Springs where the doctor took charge. Shortly she was back at the ranch in good shape.

I have no idea how many cattle he ran, but during the eighties the Musselshell roundup ranked as one of the biggest in Montana territory, and Two Dot's outfit was one of the larger ranches. What was called the upper Musselshell range then stretched from the Castles and Crazies on the west to the Big Bend of the Musselshell on the east, and from the Little Belts and Snowies on the north to the Yellowstone-Missouri divide on the south. A big scope of country and as fine a range as there was in Montana. Or anywhere else.

He ran a lot of good horses. So good, in fact, that they were a prime target for horse thieves, and it got so bad for a while that he had to dayherd them and pen them at night. His brand was easy to work over, so he lost a lot of stock, but he didn't think it was a damn bit funny when a neighbor suggested that his brand would show up better if he put it on with the bottom of a red hot skillet. He took good care of his horses, too. One bad winter, caught in a blizzard, he rode all the way home to the ranch and came within an ace of freezing to death doing it. The reason he didn't hole up somewhere along the line was that he couldn't find shelter that he figured was any good. For his horse, that is, not himself.

When Two Dot hit the Musselshell country, there were still buffalo on the range, and he must have killed a few. Anyhow, one of his prized possessions was a mattress stuffed with buffalo hair. Guests were proudly given the room where it was on the bed. Gramp told me of one time when he and Granny drew it on a visit to the Wilsons. He said there was more give to concrete and that the damn thing had a bulge in the middle so bad he had to hang on to keep from sliding off onto the floor. He wasn't alone, for he could hear Granny muttering to herself—of course, she was holding on for dear life 'way over on her side of the mattress. She got the giggles though when he remarked quietly so their hosts wouldn't hear, "Must be buffalo, sure enough, Alice. By God, it's even got the hump!" They spent the rest of the night in chairs and lied manfully in the morning.

A grizzly out of the Crazies got to killing cattle. Two Dot took it as a personal affront. So one day he saddled up, got his rifle and went bear hunting. By himself. Said he didn't need any help to down a damn bear. He was back toward evening, empty-handed. Naturally he was questioned. Had he seen the bear? Had he killed it?

"Yup," was the answer. "I coulda shot the sonofabitch."

Then, very judiciously, "But I decided I hadn't lost any grizzlies. He was too damn big to pack home anyhow."

I never heard whether they finally got the bear or not. If they did, I'd bet that the old man took along some help for that go-round.

His outfit was something of a soft touch for grub line riders, thanks to his wife. Besides, in those days if you were anywhere near a ranch anywhere around a mealtime, you stopped in. The rancher would have been insulted if you hadn't. Two Dot wasn't wholeheartedly in favor of the custom, at least when it was his outfit where the stopping in was done. It bothered his parsimonious heart. One day when Dad happened to be there at dinnertime—lunch was always "dinner" in ranching Montana—it got the best of Two Dot. There'd been a dozen or more at the table, mostly visitors, and the meal was about over when a rig rattled up to the house. The Wilsons' old dog heard it, got up from his place by the stove and headed for the door, wagging his tail enthusiastically. "Set down, goddam you," Two Dot rasped. "Don't be so damn friendly. Hell, you don't have to feed 'em. I do." Then he got up and went to the door to greet whoever it was.

There was another deal that gave him a lot of satisfaction. The woodpile for the kitchen was carefully positioned right beside the path to the outhouse and close to that edifice itself. It was handy for anyone coming back from the privy to bring along an armload for the woodbox. It worked to beat the band, I guess. Especially with the hired girls. Most of them were pretty shy and seemed to figure that if they were spotted coming up the path with stove wood, nobody'd suspect what they'd actually made the trip for. A two-way proposition, for the girls' sensibilities were protected and the wood box kept full, though nobody was fooling anyone.

The setup in those days was a lot simpler. Now you have to get a permit to put in a septic tank on your own outfit. It

can't be anywhere near running water, either. Hell, I
remember a few places that were proud as could be of the
fact that their privies were out over creeks. Kept 'em nice
and clean and fresh. Then there's all this hullabaloo lately
about how wood smoke is absolutely loaded with poison or
something. Most old-timers I knew used wood all their
lives, and by God, they didn't die very young. Maybe they
were just lucky, but it's a cinch that they were tough, and
hard work kept 'em that way.

In his later years the old fellow didn't take as active a
part in the stock work. He was getting a little heavy to do
much ahorseback, but every day his private mount, an old
white gelding, was saddled, brought over to the house and
pastured in the yard. It took the chore boy less time to
pitch out the horse manure than to cut the lawn as Berry
wanted, according to Two Dot.

One day he decided to take a trip over to where the crew
was working some stock. So out he went, bridled the horse,
tightened the cinches and stepped aboard. His pet hadn't
been ridden for quite a while, was full of oats and green
grass and be damned if he didn't blow up. He couldn't buck
hard, but he bucked loud, and Mrs. Wilson hurried out to
see what the commotion was all about. The old pony was
bucking along by the gallery of the house. To keep him
from going under it, she hurried along in front of the two,
flapping her apron in the horse's face, calling "Shoo, shoo!"
She got them headed away and then gave instructions,
"Get off, George. Get off!"

Two Dot was still in the buggy, though he was a little
loose and getting badly shaken up. When he had the time
to spare, which was between jumps, he grunted an answer
to her plea, "Get off—hell. I'm doin'—goddamned—good
to—stay on!"

But stay on he did. Even went over to the cattle work.
He didn't ride much after that, but when he did, it was on
that same horse. He was kept ready in the yard as usual,

but I imagine they cut down on his oats.

As a matter of fact, I believe the last horse to carry the Two Dot brand was Gray Dick. He had quite a reputation as a horse that was awful hard to ride, and he earned it in many a "bucking contest" where the rider could use his own rigging or any "bear trap" outfit he wanted. He was 'way past his absolute prime when Jess Coates fit the first qualified ride ever made on him years ago. Gib McFarland owned him for a long time after Two Dot was gone, and the old horse died on Gib's outfit not too far from where he'd been foaled, for he probably was kin to Two Dot's old white.

My Aunt Allie went over to visit the Wilsons and to show them her baby boy. She also took along the nurse, who was a pretty substantial chunk of woman. They were invited to spend the night and accepted. During supper Two Dot inquired idly, "Where's that baby goin' to sleep?"

My aunt told him, "With Miss Peebles."

He grunted noncommittally, and they went on eating. Pretty soon he fidgeted around in his chair and looked up. "Who'd you say was goin' to sleep with that little feller?"

"Why, Miss Peebles there," answered Aunt Allie, indicating the nurse.

Two Dot sized up the woman and went back to his plate. Something seemed to be bothering him, for he suddenly laid down his knife and fork. "Did you say that little boy was goin' to bed down with her?" pointing to the lady in question.

"Yes," was the answer. "Why?"

Two Dot could contain himself no longer. "Damn woman'll roll on him an' kill him!" He was a stockman all the way!

When the railroad came through the valley, Two Dot, like a number of other ranchers on the river, was not eager to have his bottom land messed up. So a man came around to persuade him. He found Two Dot a hard nut to crack, and what he considered one of his best arguments fell on barren ground. "Think, Mr. Wilson," he enthused, "with a railroad through here you could ship all your hay to market."

He made a bad mistake, for hay to most Montana ranchers is the hole card necessary for bad winters. He found it out right quick. The old man drew himself up in high dudgeon. "Who th' hell's peddlin' hay?"

The man persisted, and Mrs. Wilson telling Gramp about it later, said bitterly, "Why, he was so rude I almost wanted to shoot him."

Two Dot, who had been listening quietly, turned to her, "By God, Berry. You're th' killin'est damn woman I ever saw. Where do you bury your dead?"

Gramp said it fussed her pretty badly, too.

The man kept coming back, and I guess the Wilsons got tired of arguing and unlike in earlier days he couldn't just be gotten rid of. At last, he was finally told, "All right, go ahead an' build th' goddam thing."

I don't think they ever got a cent for the right of way, either. Sounds pretty typical of a railroad to me. At that, he'd have the last laugh today, for the original railroad, the "Jawbone"—named because it was supposedly built on talk—and its successor, the Milwaukee, are gone from the Musselshell. Probably for good.

There was one saving grace, at that. When the railroad came through in '99 or thereabouts, Two Dot had himself a town. He donated part of the site and took an active interest in developing the burgeoning city, particularly so since it was given the name "Two Dot," as was the post office, established in 1900. I've heard that the first building was a store. That I doubt like hell. My bet is that it was a saloon. Two Dot had no part in any of the thirst emporiums as far as I know—he left that up to Gib McFarland and Knute Hanson, who moved their outfits over from the inland settlement of Big Elk, eight miles to the south. But inside a few years he had built a big store and a hotel. He also contributed toward a school.

In most frontier towns the saloons came first, the stores and blacksmith shop second, and the churches third. In order of importance, you might say. Two Dot was no exception. As Gramp recollected years later, when they got around to the church, the Wilsons contributed heavily to it, and thereby was sowed the seed that ultimately caused him to quit being a major booster of the metropolis.

Seems that when the time came to put in the windows, Two Dot wanted his brand on them—even offered to pay for all of them himself. This infuriated a number of churchgoers. I don't know why it should have, really, for the old man had sure helped the town from its infancy. I doubt very much that God would have had any great objections either. However, good triumphed, so the windows carried nothing as crass as an old cowman's brand. Two Dot for his part decided, "To hell with th' bastards an' their town," and from then on until he died in '07, lost all interest in the civic affairs of his namesake.

Even so, though Two Dot, like Melville and a lot of other little towns in Montana, began to play out towards the second decade of the 1900s, it would warm his stockman's heart to know, wherever he is, that Two Dot, Montana, was the biggest single livestock shipping point on the Milwaukee up almost until the dissolution of the line in 1981. The latter covered a big scope of ranching country from St. Paul to Washington state. I think he'd be glad, too, that Jewell Harper is on the old ranch running it as a good cow outfit and living in the same house he lived in. It should tickle him, also, that the most thriving business left in his town is the Two Dot Bar—the first shall be last, as they say.

He was quite a guy. A sure-enough man for his time. As he once told Gramp, "Paul, I've lived my life th' way I wanted to. An' I can look any damn man in th' eye an' tell him to go to hell!"

Not many people could say that in this day and age.

===8===
BEST OF THE WEST

RODEO HAS COME a long way since I was a button, and I'm not sure it's been all to the good. Sure, the prize money is mind-boggling nowadays, but something is missing. I guess maybe it's the camaraderie that was such a big part of the game when I was young, when the contestants were all damn sure cowboys and anybody with the nerve could enter wherever and whatever he wanted. Hell, there weren't many entry fees even when I was a kid and the purse was likely as not made up by passing the hat. Yes, and if a man had bad luck, it's a cinch somebody who'd hit a lick would stake him with enough to make the next show or to get home, anyhow. That's pretty much a thing of the past.

Nor can a man just gather up some salty horses and cattle anymore and go ahead and put on a show. Nope, he won't get any contestants—everybody belongs to some rodeo association or other and can't compete unless their outfit approves. Sort of sad.

When I was a kid, if you'd said "rodeo" to somebody, he'd have wondered just what in hell you meant. Of course, since Hub Hickox founded the town along about '80, it's a cinch that every celebration—and Melville was always strong on celebrations—had a "bucking contest," a steer jerking and always a horse race or two. Or three or four. The "Fourth," the "Seventeenth of Norway" or just when some old boy brought in a horse that he claimed "can't be

101

rode." A hat would be passed, and somebody'd sure have a sitting at him. I remember many a picnic down on the Sweet Grass when I was little, and there'd be a bronc ride or two at every one of them, barring the Lutheran Ladies' Aid doings, maybe. But I wouldn't bet on even that.

From what I've heard, there were some salty riders and bucking horses back before my time. Two Texan brothers, Mel and Jase Jowell, when they weren't doing one of their many stints in the pen for stealing horses or cattle, were popping riders, though Mel was the better of the two. He also was handy with a six-shooter. Old-timers tell of a ride he fit on a grey bronc at Two Dot that was something to write home about. Charley Bennett, a New Mexico boy who'd learned to stand the winters up here, was another top rider. He went down to Big Timber in '04 and won the bucking contest there and then at Billings the same year. Considering the scope of country and the riders representing it, he must have been pretty good. Dad, as an eighteen-year-old, beat him in Big Timber in '07, but that sure was no disgrace.

Charley was quite a roper, too. Of course a "roping" meant a tie-down steer roping: big cattle, and it was about an even break who'd jerk who down. Nobody ever dreamed of roping *calves*. I've heard of one instance when Charley built to a steer without remembering to tighten his cinches. He latched it on, threw his trip, turned off and his memory came back in a hurry. They say his horse was a little owly to rope off for quite a spell after that. Scotched pretty badly. Charley must have quit punching cows, for he married a local girl and ran a saloon in Melville for some time. The same saloon where Barbara and I lived while we were building our house a quarter of a century or more later.

A number of cowpunchers turned saloon keepers in the old days if they could save up enough money or win it—or the saloon—in a card game. I've had pretty deep misgivings as to how it must have usually worked out, for most cowpunchers I've known would have been their own best customers. Once in a while a saloon man would become a rancher. Not a cowpuncher but a sure-enough cattleman because, hell, next to the bankers (and sometimes in front of them) a saloon keeper was the richest man around. I could name a couple, anyhow, who parlayed a saloon into a first-class cow outfit.

Later, when I was growing up, the Melville country and the Sweet Grass had a number of well-known bucking horse men. Sid Brannin, Johnny Christenson, Roy Connolly, Miller Pedersen, Dad and a number of others.

Sid was a booger. He liked horses that bucked and sure saw to it that they did, whether they really wanted to or not. I remember up at Billy Creek—it must have been right at sixty-five years ago—when he took a sitting at one in the big round corral on the flat south of the house. Funny what a little kid remembers. The horse was a sorrel, and Sid was wearing angora chaps. He no more than hit the cantleboard than the pony stood him right square on his head in front of the snubbing post. I can see it to this day, the post and those chaps upside down beside it, Sid in them. I whooped, "Can't you ride him, Sid?" and came within an ace of falling off the top pole, only Mother saved me.

I'm sure it would have suited Sid if I had, for he'd probably have throttled me gladly, but he never said a word. Just got back on and rode the horse to a fare-you-well. Whooping, fanning or thumbing all the way.

That "never said a word" was unusual for the man. He always made a lot of noise—he could back it up, too—and had a lot of fun. He broke horses all around the country. Trouble was, he liked excitement, and the horses got so they did, too. He could turn out a reined horse, a rope horse or anything you wanted. And nearly always, sometime during the day, you'd sure have to look at your hole card. I can remember a few of them real well, for he broke a string for our outfit one spring. Seat Up High, Trip Rope and Fizzy were three of them; and none of them ever plumb quit wanting to buck. It tickled Sid. He always had a gunsel or two that had their shingles out as cowboys working with him. None of them ever seemed to last too long, for he delighted in getting them bucked off, run away with, stomped, anything. He was a hell of a hand himself, but he sure loved to job people and had so much fun doing it that nobody got wolfy. He also was the owner and user of a particularly gaudy set of cuss words. One phrase especially fascinated me, as a button, with the picture it drew. I won't go into detail, but it featured a wildcat, a red hot awl and butter.

Johnny Christenson was a horse of a different color. Big, stout and rough, he would tackle anything that wore hair. You couldn't have dented his hide with a new rasp! Once he was riding a big rank bronc on circle—for the Triangle Anchor outfit over on American Fork, I believe it was. He dropped off to make his gather, and a while later one of the other riders noticed John's horse, standing sort of awkwardly off on the side of a hill, and loped over to have a look. Lucky he did, for the horse had turned over and hung John up. But as the bronc got to its feet, Chris had reached back and gotten a death grip around both hind legs at the hocks, and he'd held that stout booger until help came. I never heard how long that was, but he was a goner if he'd lost his grip. It took one hell of a man to do what he did! All he said when the other fellow showed up was, "Wondered how long it'd be before somebody'd notice us."

I saw him ride Man Eater to a complete finish in the corral back of our barn—the wickedest, most human-hating horse I ever saw. Never did get him broken, and he was damn dangerous around the place, so Dad sold him to Cremer for a saddle bronc. Didn't even do for that. He was so mean in the corrals they couldn't handle him, and if they got him into the chute, he fought until he played himself plumb out. Not only that but he got hold of one rider, jerked him down into the chute by his chaps and almost killed him. Bawling like a grizzly. So Cremer did away with him and shot him like the wolf he was.

Roy Connolly was another good rider. I never saw him in his prime, but he sure knew bucking horses. He was greatly responsible for that ungodly string Cremer had. "Buffalo" helped put it together. When he said a horse could buck, Leo would buy it, even if he hadn't seen it. On top of that, Roy was a first-class cowman.

Miller Pederson was a dandy, too. To see him you'd have thought he was a big old farm boy, but make no mistake, he was a bronc rider. He loved to ride a bucking horse: for prize money, a matched riding, on a bet or just for the hell of it. He figured a man should make a little noise when he fit a ride, and win, lose or draw he never quit grinning. He tackled an old pony once that had quite a reputation, just about the time that bronc riding rules were getting up into this country. Up until then, around here, a rider either bucked off or rode his horse to a finish and be damned to how long it took him.

Miller agreed to ride this booger according to the rules, and out he came—spurring, fanning and whooping all the way. When the whistle blew, he stayed with it, and when the pickup man tried to ride in, Miller told him loudly, "Git away, goddammit, git away. He's still got some buck in him." Then he proceeded to romp on that reputation horse from hell to breakfast until the old pony just flat quit. The owner was a little ringy about having his top bucking stock treated that way, and Miller couldn't understand why. He'd just ridden the horse the way a bronc ought to be ridden. What the hell. Quite a guy and a damn fine rancher, too. Over the years he's put together one of the best outfits in Sweet Grass County, and he hasn't changed a lick. Still the same happy go lucky son of a gun!

I've never been able to figure how it happened that the

Melville country hung fire until 1931 before it put on a sure-enough rodeo. Hardly in keeping with the local esprit de corps. Naturally, though, that first show was very much of a community effort. An arena of sorts was built east of the Sweet Grass back of the schoolhouse, and, with the usual civic optimism, a two-day show was scheduled.

It was pretty much of a punkin rolling. Sorry chutes, damn little organization, but all sorts of excitement and lots of fun. Johnny Browne, one of "Slick Saddle" Browne's boys, supplied the bucking horses. He got them somewhere over in the Two Dot country, I understand. At least, it was claimed that after the show was all over he headed them up the lane to the north, gave them a running start over to American Fork and hurried back to Melville before the moonshine gave out—though there wasn't much chance of the latter occurring.

They were good horses. It didn't really rain cowboys, but there was a pretty fair sprinkling. I believe that the last saddle bronc Buffalo Connolly ever rode was at that show. Or rather, didn't ride—so emphatically so that he quit for good. He was getting a little long in the tooth anyhow. Snyd McDowell was there, but he'd quit bronc riding by then, so he had the time of his life giving advice. Just exactly how each horse would buck. He'd never seen any of them before, but once in a while he hit it on the nose, and whether the rider had stayed in the buggy or not, Snyd would be able to say, "See. I told you he'd do thataway." I liked Snyd. He had a way of putting things that was magnificent. It was hotter than the hubs the first day, and he had on a heavy flannel shirt, buttoned plumb to his ears. I asked him how he could stand it, hot as it was. I've never forgotten his reply, "By Jokes, if it'll keep th' cold out, it sure oughta keep th' heat out, too, don't you reckon?" I guess it did, for he wore it both days.

It was a bucking horse show all the way. There was no 'dogging, for nobody had any horned cattle, barring some registered stuff, the owners of which would have taken a dim view of using them. Even if anybody could have thrown them. There was a calf roping, though. I forget who won it, probably a McDowell. I know it wasn't me, though I was in it.

There were horse races, too, at least three or four every day. There were big panels at the far end of the arena, away from the catch pen, which were moved by plenty of volunteers from among the spectators before each race and put back after it was over. Why nobody got crippled when a race poured into the arena, whipping over and under and pulled up short at the yon end, I don't know. There were a few wrecks, sure, but nothing that wouldn't heal up pretty fast, for there was quite a mixup each time and those that got badly outrun were the lucky ones, looked like. But it was exciting. Right down Melville's alley. I've wondered how the lady barrel racers nowadays, who don't seem to be able to stop their horses inside the arena and all by themselves, would have liked it. Probably'd have turned in their suits.

All in all, that first Melville Rodeo was such a success, as far as the whoop and holler went (which was always dear to the Melville heart) that another was put on the next year. Charley Murphy, from up the Yellowstone at Emigrant, brought the rough stock, and Dad supplied the bucking steers and calves. Unlike the first year when the contestants were almost entirely local, that year there was a sprinkling of outsiders. It was a good rodeo, but like all Melville shows, what happened outside the arena was just about as interesting as what went on in it. Maybe even more so. Over the years there were eighteen rodeos at Melville. I saw and contested in all of them but one. I missed it because Barby was born that morning, and I went to Lewistown to see her and Barbara.

Speaking of the fringe benefits, the non-arena things (though from a technical standpoint I suppose you could

say it belonged to the arena), reminds me of Charley
Hoffman in '32. He'd filled up on moon and made a damn
nuisance of himself all during the rodeo. So, figuring he'd
do the same at the dance, when everybody'd left the
grounds, a few of the more farsighted town fathers snagged
Charley and tied him in one of the bronc chutes. They
spreadeagled his arms out and over his head so he couldn't
reach the ropes with his teeth and left him. They propped
the chute gate open, though, so he wouldn't feel too
confined.

With all the festivities up town they plumb forgot about
him until morning. Then, a little apprehensively, they
hurried down to see if he was still alive. He was, and sober
for damn sure, but he was in rocky shape. The night's dew
had shrunk the ropes so they had to be cut loose. Besides,
the morning sun was blasting into his face so he was
sweating something fierce and was rumdum enough so's
when they told him they'd heard him yelling for help, he
didn't wonder why they hadn't come sooner. He'd yelled,
all right, but all he could do by then was to sort of croak
and cuss. While his saviors were getting him back to town,
somebody, sort of slaunchwise, mentioned that Sandy
Harper had been giggling around the night before, and he'd
wondered why. Charley jumped on the idea, for it was just
Sandy's caliber, and got real warlike. Of course, Harper
hadn't had a thing to do with the deal, but it was quite a
spell before he'd come into Melville after that without
scouting it pretty thoroughly first. Hoffman was an awful
good man with his hands.

The next year three ranches, the Lazy K Bar, the Dot S
Dot and the Sixty Nine, took over the show. There was no
place where spectators could spend the night except
Harlowton, twenty-five miles to the north on the Musselshell, or Big Timber, about the same distance south on
the Yellowstone, so the second day had always been pretty
sorry. Besides, there was no eating place in town—much to

the glee of the Melville store, which did a land office business on anything edible. Of course, there was plenty to drink, but man cannot live on booze alone, though some of them sure tried. Also, if the first day was Saturday with its attendant dance, the average spectator was so hung over on Sunday that he wasn't likely to get to the show that day, unless he hadn't been able to make it home. If the first day was Sunday, those who were still able to work on Monday had to, for it was haying time—the wild hay was ready.

So, since a rodeo has to have some sort of gate receipts— though I made more contesting than I ever did from my share of the gate—we cut it back to a one-day affair and permanently set the date for the second Sunday in August. That was a good date dudewise, and we incorporated a few dude events into the program. Since dude races would be a little juicy, what with the congestion at their finish, in '33 we moved the arena west of the Sweet Grass where there was more room and less chance of crippling a guest.

Everything went better over there, so after a couple of years (in one of which Cremer supplied the horses and there was a veritable cloudburst of bronc riders) we rebuilt the arena and graded a race track just south of it. The sure enough Melville Rodeo came of age that year, 1937.

We supplied all the stock. Dad had horses, God knows, so before the show we'd make a trip to the "No Bush" range, gather a string, cut out a good-sized bunch of threes and fours and trail them to the Melville ranch. The threes were barebacked, and the fours were saddle broncs. They were all hot bloods, quick, squirmy and hard to ride. Funny thing, they didn't seem to remember being contested and broke out fine later. A lot of them that fired in the arena went on to become good dude horses. At first Dad supplied the cattle, too—big, stout whitefaces. Then, in '39, we brought up around a thousand RO steers from Sonora. We dehorned all but about fifty head that had nice horns, and when we were gathering the bucking horses in "No Bush," we'd work these steers, too. Those we had trouble outrunning we threw in with the horses and trailed up with them. Hell, those corrientes traveled like ponies, and in the arena you damn sure better be ahorseback or the further they went, the dimmer they grew. I'm sure Melville had the first team tieing in Montana, for I'd seen it in Arizona when we were there the winter of '39. It was a sure-enough cowboy event, so we put it in the show that summer, using these steers.

I'd brought up a hundred or so she-stuff the same time as the steers. They cost me $27.50 a head laid down in Big Timber, including the freight, so by '40 I was supplying half-Mex calves for the ropers. We used nothing but heifers, ran them before the show but didn't tie them, and they were juicy. And then some! Dallas Brunson, a good calf roper from Billings, as I remember, took right at a minute to tie a heifer he drew. She bounced like a rubber ball, met him when he came down the rope, it was about even up who'd get who down, and then she peeled him from hat brim to boot tops. Dallas had figured he'd just flat eat up those old country boys at Melville, too. Nope, he wasn't jobbed. He drew her, and she wasn't a bit ranker than the others. Only he hadn't roped that caliber calves before. I doubt he did again, for he didn't come back the next year.

We had a number of entrants that went on to be hellish hands on the RCA circuit. Scrapiron Patch from down at Miles City ended up one whale of a bull rider. Lup Linderman won the bareback one year. He died not long afterwards, but I've been told by men who knew that if he'd lived, he'd sure have hurried his brothers. Bill Linderman blew a stirrup at Melville, back when he was starting, and didn't buck off only by the grace of Lady Luck. When he was World's Champion Bronc Rider later on, I claimed a dude had taken a movie of that ride and would threaten to show it. He'd just grin. That was Bill. A nice guy and a hell of a hand. He cashed in like the man he was—trying to help someone else. I liked Bill and am proud to say he was my friend.

There was an Indian boy from Alberta (Cardston, I think) that used to come down. Jimmy Wells was his name. He was quite a bronc rider, and later on I heard he won some Canadian championship—either saddle bronc or bareback, I believe.

There were others, like Bob Shelhamer, who never quit ranching to hit the circuit. If they had, they'd have done well. Bob was a rough all-around hand: saddle, bareback, roping, you name it. Actually, we had so many entries that in an effort to cut them down, we kept raising the entry fees. The purses, too, naturally, but not to the same extent. Didn't do a bit of good, the number of contestants kept climbing each year. It was an amateur show, a cowboy show, all the way. 'Dogging wasn't connected with range work, so we didn't have it. Instead, we had team tieing, and for a couple of years, steer jerking. The SPCA took a dim view of that, besides it was illegal in Montana at the time, so we quit it before we got in trouble.

There was no catch-as-catch-can roping, either. We figured that if an old boy was roping at the head, then, by God, he better catch the head. First, anyhow. In all the ropings a man could pack only one rope, because, hell, a working hand never carried an extra catch rope with him on the range. Me now, I never knew whether the one I did have with me was a catch rope or a throw rope until I tried it out. Pretty often it was the latter rather than the former. I don't know whether there were bareback riggings back then, but there sure weren't at Melville. They rode those colts with a loose rope, and I can imagine the reactions of today's bareback hands if they had to do that. So our rules differed from RCA rules on a few points, but not much.

Barbara and I were the secretaries. At first, we took entries at the gate to the grounds, but after a year of that

we moved up town to the porch of the Bucket of Blood—
the Melville saloon, by a more dignified name, though I'm
not sure that "dignified" is exactly the word to use in
conjunction with the Melville saloon. Besides, it was
handy to the telephone in the store, the only one in town.
Our rules were posted right there, and nobody could enter
until he'd read them. If he signed up, come hell or high
water, he abided by them. There were a number of
instances when some old boy, usually a gunsel with his
shingle out as a sure-enough rodeo hand, would complain
about them. I had a stock answer, "Buy yourself a ticket
and watch then or go back to wherever you came from.
Doesn't matter a damn to me, but if you rodeo here, buster,
you do it under those rules." It took about one treatment
for them to get the message that we meant it.

The rodeo secretaries nowadays have the world by the
tail, and a downhill pull for entries are chopped off two and
three days, at least, before a show. We signed 'em up until
one o'clock, and the rodeo started at two, and we used to
have at least a hundred contestants. It was sort of fun, at
that, and the phone calls I got tickled the hell out of me.
They came from Judith Gap, Greycliff, Rapelje, Spring-
dale, Harlowton, even Billings. Telling of flats, breakdowns
and gas tanks running dry, along with pitiful pleas to
"enter me" or "hold my stock 'til I get there, can you,
Spike?" I always did—you can't blame a guy for a little
tough luck. I remember one old car full of riders coming
into the grounds on a rim. They'd had to pay at the gate to
get in, for they were late, but all were packing broad
grins—they'd made it!

I never even gave a thought to the chance that I might
get left holding the sack if the contestant whose entry I'd
paid didn't win anything. Of all the boys I staked, Buck
Ahern was the only one who welched on his word. He still
owes me a saddle bronc entry. All the rest paid up. Might
take some time, but I got my money. A time or two Joe and

Albert LaFountain from over in the Lewistown country came up short on their entries, but, by God, a month or two —or six—later, I'd get a letter from them with my money in it. Usually in dollar bills. If they'd overpaid me a little, I'd knock it off when they arrived the next year. They were good guys, Joe and Albert. I liked them. Also, when they showed up, I could pretty well figure on bucking off anyhow two riders. I never could understand why they kept coming, for our horses sure had their numbers. I guess it was just for the fun of rodeoing, and their rides were spectacular, to say the least.

This is bucking horse country, so naturally the saddle and bareback fees were the highest, as well as their prize money. We added the fees, and to start with, paid fifty, thirty and twenty percent of the pot for first, second and third respectively. Later, when we had so damn many entrants and high fees, we changed it to forty, thirty, twenty and ten percent, for there was a lot of money in each event. Buck Cheney of Belgrade and Hup Davis of Bozeman hit a lick in the ropings one year and took home right at $800 between them. Hell, you could win considerably more money at Melville than at the big one-day professional show at Big Timber. Hugh Stewart of Martinsdale was a regular participant. He'd ridden for the CBCs, and to make a hand with that outfit you had to be pretty salty. He was! He won the saddle and bareback both one year, and I saw him up in the saloon later with quite a tail. Hughie was a friend of mine so I walked up and said, "Hugh, you hit a lick today. Could I borrow some money?"

He sized me up, a little blearily, "Hell, yes. How much you want?"

I did a little figuring. "Could I get about $400? I need it bad."

"Well, let's see." He went through himself, started pulling out bills and counting. "Yup, I got it," and he handed it over.

I thanked him, went down home, put a rubber band around the money and cached it in my bureau. I didn't see Hugh again until in the winter when he dropped over for a stay with Cliff Reynolds. I ran into him up town, we visited a little and I asked him down to supper. Barbara was glad to see him, and while we were eating, I said, "I've got something of yours, Hughie," went in, got the money and gave it to him.

"Where th' hell'd this come from?" he asked. I told him, and he grinned. "Thought I'd spent a lot that night, but I was celebratin'. Damn, I can sure use this a lot better now than then. Thanks. I'd 'a blown her all."

One time Hugh and I rodeoed down in Roundup, almost forty-five years ago. We rolled our beds down at the stockyards. Word must have gotten around amongst the mosquitoes that we were there, for it looked like they'd come for miles, both up and down the river, to favor us. After one night trying to sleep with the tarp pulled over our heads in the middle of July, we came up with a solution: maybe alcohol would work. It did, taken internally. Didn't faze the damn mosquitoes, but at least we couldn't feel them though we both looked like we had leprosy by morning.

Roy McCaffery, from Musselshell, put on the shindig. Roy reminded me of Dad. Pretty wild. So it was quite a show. Short on equipment, so Roy borrowed a cinch for one of the Associations from a local boy who was in the roping. After the show I happened to be there when he returned it, and the old kid he'd borrowed it from was pretty owly about a few little fuzzy spots where spurs had hit it.

"Hell, it ain't hurt," Roy assured him.

"But I got a lot of big stuff to rope."

"Sure enough," Roy came back, just busting with curiosity. "Who d' you get to ketch 'em for you?"

It was a bucking horse deal from away back. They told

me that the calf roping was the second event. It was. I got ready and watched while they bucked out everything buckable, rerides and all. So, time we roped, it was not only getting towards sundown, but also there were so many people in the arena that a man had trouble seeing his calf. Sort of had to rope at the swirl in the crowd. Anyhow, I won it, and Hugh the saddle bronc. I forget what he made, though I sure remember my winnings. They weren't top heavy—a single Navajo saddle-blanket, a Stetson hat, two pair of Levis and $27.50. But we had a hell of a lot of fun. So did the mosquitoes.

But to get back to Melville. Our regular events were saddle bronc, bareback, calf roping, team tieing, wild cow milking—and we always threw a few steers in with the big rank whitefaces we milked—a relay race, a cowboy race and a free-for-all race. It's a shame you never see a relay race any more, for it was one of the finest, most exciting events at a show. Dad or Tor Anderson, whichever was riding Dad's string, was never beaten at Melville, though Bunny Beley crowded them badly a time or two. The only qualifications for the cowboy race were that a stock saddle and curb bit must be used, the weight must be two hundred pounds all told and the rider could have no help handling his horse. The free-for-all was just that—no limitations as regard horse, rigging or weight—and if it took half a dozen men to get him out on the track for the start, what the hell. You had to damn sure be ahorseback, though, if you figured on winning a race. A number of entrants found that out.

We also had a dude boys' race, a dude girls' race, a dude steer riding and a dude calf roping. The latter was one of the funniest deals I ever saw, particularly if somebody happened to catch a calf. We didn't need a clown.

For a while Sandy Harper was a sort of self-appointed announcer. He came up with a megaphone from some-where and would ride around the arena bellowing pronouncements, most of which were wrong. He was a pest, but he had so much fun that we figured, hell, let him do it. This imitation of Fog Horn Clancy came to a sudden halt at one of the shows. Sandy was out sounding off in front of the chutes, making so much noise I guess he didn't hear the warning he was given, when they turned out a big saddle bronc. Harper was in the wrong place at the right time. That bronc hit his old white horse square amidships and flattened him, Sandy on the bottom. It was a wreck! He rolled them like a barrel, bucking on top of them all the way, his rider not missing a lick. When he finally turned them loose, we rescued Harper and found he wasn't hurt. Only badly shook up—he'd been just drunk enough so's it had saved him. Anyhow, he sure quit the announcing business. His megaphone was mashed, too. We used a sound outfit from then on. Nice change.

The contestants were all working cowboys and damn good hands. Corny Pernot, Eli Chase, Speck Cockran, Albert Vermandel, Pete Harms, Jesse and Bob Langston, Joe France, Bud Reynolds, Bill Randall, Larry Chapel, Johnny Branger and Speed Komarek were a few that stick in my mind from the riding events. Aubry and Mabry McDowell, Duffy Crabtree, Don Slade, Floyd and Bill Hicks, Larry Jordan, Don Sturgeon, Jay Parsons, George Komarek, the Anderson boys (Kermit, Leif, and Tor), Jake Frank, Allan Woosley, the Bohleen boys, Bill Stovall and a lot of the riders were among the ropers. Seems to me that I've heard that a Stovall had a hell of a fight with some old boy in Billings. The event took place in a revolving door! Could have been Bill, though he was a good-natured cuss. He and I got into trouble with Albert Vermandel at a Wilsall show. Jesse Langston was in the service, and Albert had his saddle. He jerked the horn plumb off it in the team tieing and left it with Bill and me for safekeeping while he was riding his saddle and bareback horses. We swapped it to the beer man for a six-pack, as a joke. Albert's sense of humor was always pretty brittle, so much so that we went down and got the horn back. Right now.

Albert was a hard man to whip. He and somebody, whose name I can't remember offhand, had a falling out in the Grand Hotel bar in Billings. They foregathered downstairs in the men's room. I be damned if I know who won, but time they finished there was blood all over the place. I understand the Grand asked them to take their business somewheres else after that.

Hell, there's no way I can remember all the boys that contested at Melville over the years. They came from most of Montana, northern Wyoming, southern Alberta and once in a while from western Dakota. They were a good outfit. I don't remember an instance when somebody was going to "whip the judge." That used to be pretty common back in the twenties. Our judges were all men who savvied bucking horses. Might have been, too, that most of them were pretty handy in any sort of squabble, and everybody knew it.

Oh, I don't claim there were no fisticuffs at Melville. Not by a long shot! Sometimes they started before the show, even. We were running the cattle the evening before one rodeo using breakaway hondas. When we finished the steers, we started on the calves. Bob Langston backed into the box, came out, warmed his calf's britches and about fifty yards up the arena whipped it on. His horse hadn't been showing too much of a whoa, but Bob quit him as he jerked the slack. There was a pop like a .22 shot and down he went. He didn't get up, so we hurried out to see what the trouble was. "Think I've broke my damn leg," he told us, and he sure had. That's what we'd heard.

Barbara loped over to get a car, some shingles, a pillow and bandages, while we straightened Bob out and cut off his Levi leg. There were a couple of contestants who'd been at Melville before, camped on the Sweet Grass behind the arena. They saw the commotion and strolled over to watch. Or maybe to help. Anyhow, they'd either been patronizing the Bucket of Blood pretty heavy or they'd brought their own jug from home, for they were feeling damn little pain.

That didn't last long. They kept pushing in, getting in our way, all with a lot of talk about how much they thought of "Ol' Bob" until Jesse Langston, Bob's cousin, lowered the boom on the youngest, and biggest, of the two. I'll bet he knocked him thirty feet, but up the guy got, and the two built to it. It was sort of funny, peculiar, that is—there we were, busy splinting a broken leg, while the fight went on around us in a cloud of dust. It would have gone over us, Bob included, but when they'd get too close, we'd get up and point them in another direction. We had our patient in pretty fair shape, including a sizeable snort of booze which Barbara had brought back, by the time Jesse had his man whipped. Bob was taken in to Doc Baskett in Big Timber.

The next day he was back watching the rodeo and claiming we all were damn lucky he couldn't compete. Jesse's sparring partner, pretty much the worse for wear (which is a definite understatement) and his friend (who sure hadn't wanted any part of the excitement) packed up their camp and went home. They never came to a Melville rodeo again. The rest of us went back to running calves until we'd worked through them all.

There never was much trouble during the rodeo. The thirty feet of hard twist, coiled tight, that the arena director and the pickup men carried could have had some bearing on that, but there was one glaring exception.

A rancher who'd been feuding with a neighbor got a few under his belt, saw the guy sitting on the fence and jerked him down off it. The hell of it was that the jerking was *into* the arena. If it had been done to the outside, it might not have affected the rodeo to the extent it did. Anyhow, this feud was pretty common knowledge, both sides had backers, so in no time most of the spectators were in the arena where they could get a close up view of the action. Even some of the women, those that didn't have kids to keep an eye on. The arena was packed and, naturally, the show came to a complete halt. Barbara had the idea of turning out a loose bronc or two, which might have broken things up, but our chute crew was off watching, too. I was up above the chutes—announcing when I wasn't competing—where there was a dandy view, so I decided, for the nonce, just to relax and enjoy it.

It was a good fight. The crowd was so thick that neither combatant could turn it into a foot race, so the action was brisk. Wasn't long, though, until it began to look like the jerker had bitten off all he could handle. Just about then one of his friends crowded loudly into the circle. He made a bad mistake. The jerkee's wife, who was built like a brick outhouse and just about as stout, had gotten a length of two-by-four somewhere and was standing guard to see that her husband wasn't doubled up on.

Naturally, when this individual burst into the circle, she took no chances and swung her club across the back of his head. From the ground up, with both hands. From where I sat it sounded just like hitting a quarter of beef with the flat of a double-bit axe. Damn wonder it didn't kill him. What saved him, without any question, was the fact that he was drunk and a purebred Norwegian, to boot. A Norsk is hard to kill, and a drunk one about impossible. Hell, you could shoot him through the heart with a 30-30, and he wouldn't die 'til he sobered up, but this one sure lost all interest in the proceedings.

A couple of Samaritans got the cadaver by the legs so neither of the battlers would trip over it, pulled it through the crowd and laid it out by the fence. Then they hurried back to watch. Watching was about all anybody did from then on out and pretty discreetly at that, as long as the lady and her two-by-four were poised and ready.

Finally, the original match broke up. By mutual agreement, I'd say, for neither was what you'd call whipped. Sort of a Mexican standoff. But by then there were fights all over the arena. Funny thing, but I've noticed that people seem to get bristly around the shoulders when they are watching any sort of a bickering. A little push or an elbow in the ribs so they can see better, and, kerwow, somebody gets swatted and away a couple of old boys will go. Hell, I've seen two good friends watching a battle all of a sudden get into one themselves. And neither knows why. Fights just seem to breed fights, that's all. It sure happened that day. Time the first trouble ended, there were at least half a dozen more in various stages of progress. One I remember particularly. A big local boy with quite a rep chose a saddle bronc rider from up in the Augusta country. I don't remember his name. He was outweighed by at least twenty pounds, but he was a regular buzz saw. The way he smothered the troublemaker with cleverness and fists was something to behold. Lord, how he sacked him out. Talk about a scrapper!

I was enjoying the action when Dad rode up, "Get your horse and rope. We'll see if we can break this thing up and get on with the show."

I did, and we, along with the pickup men and some mounted volunteers, started clearing the arena. The sheriff showed up from somewhere, and since a local man was well worked over and bloody, he was fixing to arrest the boy from Augusta. We rode up around him, asked where in hell the law had been when it was needed and let him know that, by God, if he took somebody in who had won a fight but not started it, he'd have to arrest us too. He was a little huffy at first but reconsidered after sizing us all up. So we finished clearing the grounds, and the show went on with no more incidents.

Things carried over to the dance that night, to judge from the number of altercations. During one of them out in the street in front of the saloon an old rancher from east of town got swatted. I don't know why. Maybe he talked when he should have kept quiet, or somebody'd plastered him by mistake. Anyway, it knocked his false teeth out, and they disappeared into the crowd. The next morning he had every kid in town out in the street hunting for the teeth, and by God, they found them! They were in good shape, barring dirt and a little manure, so he washed them in some whiskey, installed them, and went home as happy as a clam.

I got into trouble on account of a fracas at a rodeo dance in Melville one time. An old boy and I had a falling out about something and were having a go at settling it. Whatever it was. There was a guest from the dude ranch, a tall man, who was always as busy as a pup in a 'dog town, particularly if it wasn't any of his damn business. I didn't see it, but I guess he crowded into the ring around the fight and reached for me. A friend of mine, Tige Sedgwick, figured he had no business there, so he blasted him.

The guy vacated right now, but up at the ranch the next morning he showed up at breakfast with a doozy of a shiner. The hell of it was that he claimed I'd given it to him. I was sure persona non grata around the outfit for quite a while—hanging one on a guest is not the thing to do. Dad spoke to me right strongly on the subject. Hell, I'd had my hands too full with one man, to say nothing of hitting another, but nobody believed me but Barbara. Matter of fact, I didn't know who the culprit was until I ran into Tige some time later. I was relieved but curious, for Tige is pretty stocky. "How in the hell did you reach all the way up to his eye?" I wondered. "I jumped," Tige told me with a grin.

There was another time I got into trouble at the rodeo, and Barbara sure didn't back me up. Which is a howling understatement. We'd finished paying off the winners up at the store, for the saloon was just too noisy. Barbara had taken the books and gone out to our car. I followed a minute later, and just as I was crossing in front of the car, I ran face to face with a girl I knew from Billings. She must have just come from the Bucket of Blood, for she gave a squeal, "Spike, I just loved the rodeo," enveloped me and kissed me very, very thoroughly. Hell, you couldn't have gotten a cigarette paper between us. Under different circumstances, chances are I'd have enjoyed it, but over her shoulder, not ten feet away, I could see my wife. Watching. I didn't need to see her eyes clearly to know what they looked like. Talk about a man getting chilled!

Ignoring the smart cracks from the guys on the store porch, I hurriedly peeled the girl loose, told her it was sure nice to see her again—which, all things considered, was the biggest lie I've ever been guilty of—said I had some pressing business, got in the car and got the hell out of there. I didn't get back up town for any of the evening's festivities either. Dammit, I hardly knew the girl. She was just being friendly. Barbara didn't buy that. Not by a damn sight, though it was absolutely true, and it was quite a while before I could let my whole weight down. Even now, when I am in bad odor, the incident seems to have a way of surfacing.

During the war we continued with the show, but there were changes. We put in a beer concession. It made money, but it was a damn nuisance. When the crowd brought their own booze, it was in more or less limited amounts, but with beer handy they could have all they could buy. Which was a lot. So we started to have more and more trouble at the grounds. It just wasn't as easy-going and friendly as before. There was one feature of those years that warmed me, though. Along about the time of every Melville rodeo I'd get letters from guys in the service. From everywhere, places I'd never even heard of. Letters from former contestants or spectators telling how much fun they'd had at the show, how they wished they were there with everybody and wishing us all luck. I remember one from Johnny Branger that I read over the sound system. Then there were the names I'd read in the papers or heard about every now and then—guys that would never be back to watch or contest again. I'd read those names during the show, and we'd have a minute or two of quiet, remembering. With the Crazies lifting into the sky above the tumbled hills, the Sweet Grass singing under its cottonwoods and perhaps the whinny of a bronc or the bawl of a steer from the pens. Those guys knew it all so well—the sounds, the look of it, the smells. They'd loved it, and, by God, they'd given their lives to keep it like it was.

After the war it got pretty tough to put on a rodeo. With the Remount all folded up, Dad cut 'way down on his horses. Cattle were worth too much to chouse hell out of them, the Mex steers were pretty near all gone, so finding stock got to be a problem. The state of Montana had wised up that there were gate receipts and sent out a bunch of men that reminded me of vultures or undertakers, the way they hovered around and pored over our books. We couldn't just put on a show by ourselves without somebody telling us what we could and couldn't do. It wasn't fun any more, so we said to hell with it.

After an empty year, in '49 Allan Woosley and I put on a show at the grounds. Allan had put together a string of bucking horses which he trailed over the mountains from his outfit at Wilsall, and I supplied the cattle. It was a good rodeo, but somehow lacked the traditional joie de vivre—even among the contestants. That was the last of the Melville rodeos, for Bob Hart of the Dot S Dot was killed that fall, and I moved up on Otter Creek the next spring. The grounds sat empty, getting sadder and sadder as they were robbed of planks, poles and wire for other uses. Finally, they collapsed and rotted. Today there's no sign of either arena or track except for a few shards of timber.

The Melville rodeo. It had a reputation! Still has, for

wherever the old stove-up hands gather up here—at a
rodeo, a saloon or just when they happen to bunch up on
the street—someone is sure to mention it. There'll be grins,
and the "D'y remembers" will take over. Names, incidents,
bucking horses, running horses, who won what, times, and
when the group breaks up it'll be like the show was just
yesterday.

Not only up here either. A couple of years ago I was
down at the National Finals Rodeo, the absolute best in
modern rodeo. After the performance I ran into an old
friend. I hadn't seen him since before the war when he used
to contest at Melville. When we finally caught up with
about forty years, I mentioned what a good show the NFR
was.

"Sure is," he agreed, "Finest 'doggin', ropin', an' bull
ridin' I ever saw. But Spike, they're damn short of what
you an' I'd call buckin' horses. The boys are good, hell, but
a feller can't tell how good. Shame they haven't got th'
horses to prove it on. 'Nother thing, th' fun we used to have
seems to be missin'. Too damn serious. Too much money, I
guess." He grinned, "Goddam, I wish there were still shows
like that ol' Melville doin's. They just don't have 'em any
more. She was a wild one, an' fun 'til hell couldn't hold it.
Th' best, by God, in th' West!"

═══9═══
A HORSE TALE

I HAVE OFTEN wondered why horses are never mentioned in the stories of the first Christmas. Sheep, goats, cattle, donkeys, even camels are, but not horses. Perhaps, I thought, there were few horsemen in Israel. Perhaps they were not a riding people, for horseback men have always been a conquering breed and, by and large, the people of Israel were not. However, I got my answer one cold December night from a conversation I overheard in the corrals, for I talk horse a little. Ah, you bet, a horse was there that first Christmas night!

Stars glittered cold and steely in the dome of the winter sky; blue-white, brittle, with a sharpness to their winking that was like small icicles shattering, tinkling. Dry snow creaked as the weanling colts huddled together, muzzles frosted, hair thick and long.

The trim dun filly, Trill, nosed the shoulder of the sorrel saddle horse in the center of the group. "Sam, what did he mean tonight when he grained us? Usually he says 'Good night.' Tonight he said 'Merry Christmas.' What's that?"

The silvery hairs of age above the wise eyes caught the starlight as Sam Hill turned his head. "I wondered, my dear, if any of you youngsters noticed. I asked the same question when I was your age. Here is the story I was told. Remember it always, for it belongs to us, to all horses. Our secret and our pride."

125

An owl hooted, the sound echoing hollowly among the bare cottonwoods of the creek bottom, and a shooting star burned briefly above the white peaks lifting to the west. Hoofs squealed on snow as the colts moved closer around their elder.

"We horses," began the old sorrel, "have been with man a long, long time. We've carried him and his belongings. We've pulled his travois, his carts, his wagons, his sleds and even certain of his boats."

"Excuse me, Sam," interrupted a blocky roan filly. "How could we pull boats? We can swim, but . . ."

"No, Dimity, I didn't mean by swimming. These boats are in ditches. Like big irrigation ditches. They are called canals. There is a roadway beside them, and teams like Bullet and Babe pull the boats from there."

"Oh, I wondered. You know all sorts of things, and I was sure you could explain."

"Yes," continued the old gelding, "we have carried his fighting men, pulled his chariots and guns in war and have died with him there. We supplied him with the strength to raise food, to build cities and to forge nations. Horse tracks mark his history, and the thud of our hooves has sounded down through the years of his growth. But today there aren't many of us left. Man doesn't seem to need us any more, except perhaps for pleasure or a place like this. You are fortunate to have been foaled on a ranch, a spot where our intelligence, handiness and heart are needed, appreciated and used, rather than the mindless strength of a machine.

"Possibly man has forgotten what we have done for him. We horses haven't. We take pride in it. Especially in what a horse did for one man, a new-dropped man foal and his dam. They have never forgotten and have given in return a gift to all horses, but to nothing else that lives and dies under the sky we see tonight. A gift for a gift. It happened a long time ago on a winter's night in the bleak hills of Judea."

The cavalryman was tired, cold and dirty. But, he reflected, such was the normal state of a Roman fighting man in this foul country: either cold or stiflingly hot. The sun burning down through harness and tunic 'til a man's sweat ran black with the roiling dust from underfoot; or this accursed bitter wind whipping dirt from the raw hills, driving grit into eyes, throat and nose until each breath was an effort, and the underside of a breastplate scored a man's hide like a rasp. A sorry country and a sorry, sour people, hating each other nearly as much as they hated Romans. A country not worth bothering with, really.

But it wasn't his business, he just followed orders; and, thank the gods, there wouldn't be any orders for a while. His leave had been overdue, especially after a tour of duty against those wild riders of the eastern desert. They were fighting men! Tough as their own thorn brush and, like it burning, flaring hotly in battle. The heat died fast, though, when matched with cold Roman discipline. But their horses! The best he'd ever seen, and none better than the grey mare he'd taken in that bickering last month. How she had fought for her fallen rider, and that last effort of the dying man to stroke the slim ankle beside his head as she stood over him! What was it the interpreter had said? That she was from the Nejd? That was it—the breed known as Drinkers of the Wind. Well, the gods know there's enough of it to drink right now, lass, and his rough hand touched the sweat-encrusted withers, oddly gentle.

He pulled his cloak tighter around his shoulders and wondered, yet once more, what in Hades had an old Jew peddler been doing with a cloak like this—fine, thick wool, lined with silk from far Cathay. Probably stolen. Anyhow, it was liberated now, and he smiled to himself. What more could a soldier ask after a rough duty than a fine horse, a warm cloak, pay in his tunic and a leave in Jerusalem? A good, wild town, that. And the women!

As the light failed, the wind increased until, squinting

under his helmet brim, the rider could barely make out a huddle of buildings on a desolate hillside. He rode toward them, catching as he did so a hint of movement on the lee of the hill above. A sheepherder's camp, evidently, for as the mare trudged past he saw a small band of ragged goats and sheep standing compactly, their heads sheltered beneath each other's bellies. Beyond them was a fire, its tattered, wind-torn tongues fitfully showing human figures in the shelter of the rocks. A dog slid like a shadow through the dark and dust toward the mare's heels. Her ears flattened, a hind hoof lashed out and with a yelp the dog hobbled into the gloom on three legs. The cavalryman grinned and rode on.

Reining up before what, as far as he could see, was the largest of the buildings, the soldier dismounted stiffly and hammered on the door with his sword hilt until it swung open. A figure peered at him in the dim light from inside. Yes, an inn, but it was full. But then, leaning forward for a closer inspection; a soldier, a soldier of great Rome? Why, of course, room would be made. Shelter for the horse? Well, yes, there was a rough stable, some grain, cut grass and straw. The horse would be tended while wine and food were made ready. If only he had known the great lord was coming. Oh, very well, if he wished to care for the animal himself, just follow the boy.

The stable, nothing more than an open leanto backed into the wind, was already occupied. An ass and a bony cow dozed in the front end, and in one corner of the back a man and woman, country people from the look of them, sat by a smoky clay lamp. In the other corner was an empty stall.

Stripping the mare and watering her at a trough by the front, the soldier rubbed her down and scraped away the sweat marks with handfuls of straw. He put her in the stall, bedded it with more straw, fed her grain and some hay from the corner where the man and woman sat and strode toward the entrance. Then he hesitated, took off his cloak, went back and spread it carefully over the mare's back. He turned to the couple in the corner, "Keep an eye on my mare for me, and be damned careful of that lamp."

The mare daintily ate her grain, nosed the hay until she found some to her liking and munched quietly. When she had enough she sighed, shifted her weight and dozed, warm, well-fed and comfortable. Suddenly, a soft sound penetrated her reverie, and her ears pricked. It came again, a low moan. She craned to see over the side of the stall, but the flame of the dim light was blocked by a kneeling figure. Then a warm scent filled the cold air, and her fine eyes widened as memory came.

The storm-whipped wadi. No shelter, no escape from the bitter wind and the lash of driven sand. It was then and there that she had foaled her first, and only, foal. Alone, for the men and horses were harrying Roman pursuers away from the trail, and the women and children were hidden on the mountain. Alone she had dropped a fine horse colt, and alone she had watched it die. Her busy tongue had not been able to warm the small body enough to keep it alive through the vicious night, though she had licked until her tongue was bleeding from the sand driven into the silky hair. Nosing the little form had brought no result. Nor had the gentle pawing to which, in desperation, she had at last resorted. A night like this. A tiny, stiff, chill body in the grey morning. Her foal. She shivered under the heavy cloak.

The woman took her newborn son from the hands of the man, dried him, wrapped him in her flimsy shawl and cradled him tightly against her breast. "Oh, Joseph," she said wearily, "if only we can keep him warm until morning."

There was a soft whicker. Something touched her. She turned her head. The corner of a wool cloak, warm and redolent of the clean scent of horse, rested on her shoulder. Her eyes followed its length to a grey nose above the side of the stall, to the teeth which held the heavy folds and the anxious, lovely eyes above. Gratefully she caught the garment as it dropped, and the dingy stable grew bright with her smile. "Thank you, horse. Oh, thank you."

The small bay filly, Winken, blew her nose softly and sighed. "That was a lovely story, Sam. Did the little man foal live?"

"Indeed he did," the old gelding answered. "His bloodline goes on today."

"But, Sam," asked Derringdo, the cocky, brown colt, "what about . . .?"

"Hush, now, youngster. You will learn what it is, soon. Be quiet and think of the story you have just heard. All horses everywhere have received the gift on this one night each year. Every year, from that night the man foal was dropped, and they will until the last horse is gone. Hark and watch."

Imperceptibly the night brightened. The stars paled before a golden glow which grew to fill the vault of sky, breathtaking in its warmth and loveliness. A coyote yapped querulously from a distant snow-covered rim, and the colts moved uneasily. Then, from everywhere and nowhere it came; the music.

It sang: triumphant, yet gentle; happy, yet sonorous. It carried the jingle of trace chains, the chime of sleigh bells; the lilt of bugles, the whinny of trumpets; the rhythmic song of a jerk line, the tinkle of pack bells; the boom of kettle drums and the beat of countless hooves; a stallion's challenge, the friendly nickers of companions and the shrill call of a colt to its dam. The blue dome of the sky pulsed with it. Then it softened; comforting, reassuring as the answer of a mare to her foal. And as it died, a soft, loving voice filled the night: "Thank you, O Horse. Always."

Sam Hill lifted his head proudly. "That is the gift given to us, my children, for only we horses have ever heard it. Or ever shall."

=10=
TWENTY DOLLARS FOR THE WHISKEY

I'VE BEEN TO a lot of horse sales over the years. Mostly to watch, sometimes to buy and once in a while I had stock to sell. Without exception, I enjoyed them; even when I didn't make much money, for horse sales are a great place to foregather with friends, hoist a few and invariably there is some sort of excitement during the deal. I think, though, that the damndest horse sale I ever ran into was back in the forties, up on the highline.

Matter of fact, it was the very first time I ever sold horses through a sale ring. Secondly, there were two other horse breeders involved in it with me. Thirdly—well, I guess I better tell the story.

Times weren't real gaudy back then, and not many people were coming out to the ranches to buy horses. So, and I don't remember exactly how, three of us hooked up and decided to hold a sale of our own. One was a breeder from eastern Montana down on the lower Yellowstone, who was going out of the horse business, and the other was an old Texan who, somehow or other, had a string of horses up here and no place to run them. We all had damn good horses, and it was sure worth a try, anyhow. So we made a deal for the use of one of the sale rings up north, contracted Gibby Gilbert to cry it for a percentage and got hold of Bill Hagen of "Bit and Spur" in Billings. He tended to our advertising and catalogs.

The boy from down the river had nineteen head of

beautifully bred stuff, including his stud and some good brood mares, that he fit at home. I had an even dozen popping nice youngsters to go, and the Texan, J. D., had twenty-nine head. Mostly two's and three's, some brood mares and a few using boogers. Every horse we had was a good one, all registered. I fit mine at home, and because I had plenty of corral room and grass and because the old fellow didn't have a place to go, I invited J. D. to bring his string to my place.

We enjoyed having him there; he was a great storyteller, a nice guy, and I picked up a few wrinkles from him about handling rank horses. He was a funny old fellow, though. He had a bronc tied in the barn a few days before we loaded out for the sale. Buckshot, who was about three then, figured the animal looked like her mare, Micky, and toddled in alongside. It spooked the hell out of me, Barbara and the old fellow, but we talked her out quietly with the bronc shivering, snorting and crowding over against his side of the stall. As soon as Buck was out, Barbara, predictably, snatched her and really warmed her britches. J. D. didn't cotton to that and said, "By God, any woman'd whup a little baby like that, 'specially when everything's over, why I won't stay in her house." So he stayed in Big Timber for the next couple of days and commuted to the ranch until we loaded our trucks and headed north.

The old boy from down on the Yellowstone and I were rank amateurs at putting on a sale, but not our partner. He'd been there and back! The first hint we got of it was when, the day before the sale, after we'd all gotten our stock cleaned up and looking good, J. D. collared us and said, "We'll just let these ol' boys look everything over this afternoon. Us, we got work to do."

Seeing as how everything was all shaped up, I asked, "What do you mean?" "Son," was the answer, "jest do as I say. First things first. I need twenty dollars from each of you. Ne' mine why.".

"Whoa, now," our third partner broke in. "Damn if I fork over that sort of dough unless I know how come."

"The whiskey, dammit, the whiskey. I figure it'll cost us twenty apiece for the whiskey. An' a few other things."

So we peeled off what he wanted and he left with, "Come up to my room in about a hour."

When the two of us got up to J. D.'s room on schedule, we were curious as all hell but totally unprepared for what we found. If it hadn't been for the stack of cases of Coke in one corner and a smaller pile of cases of empties in another, it'd have reminded me of a moonshiner's outfit up on the head of American Fork, there was so much booze stacked in a third corner.

"Figure I got everything we need," our partner announced. "Now watch," and he uncapped a bottle of Coke. Carefully, he poured a good fourth of it into an empty bottle. Then he filled the original bottle with whiskey, rummaged around and produced a capping machine and a box of caps and proceeded to recap the bottle of mixture. After about three go-rounds, he sized up the bottle he'd been pouring the excess Coke into, topped it off with whiskey and capped it, too. "Come on, now. You fellers got the idea. Get to work."

We did. All afternoon we mixed Coke and booze while he explained his campaign. "They's nothin' crooked about th' deal. Jest goin' to warm up the buyers. We're sellin' a damn fine string, an' this'll maybe he'p 'em bring what they're worth. She's worked for me before. Jest watch the bidders limber up."

"Course," he went on, "we got to mix one batch pretty light. For th' ringmen an' Gibby. None of 'em are wise to th' deal, an' we damn sure don't want any o' *them* to get drunk. Got to leave anyhow a case of pure pop for th' kids, too."

"What kids?" I asked.

"Oh, I got me half a dozen kids to pack buckets of ice an'

our sody pop through th' stands an' give it away. Free
gratis. We're givin' 'em a dollar apiece an' all the Coke they
can drink. But we damn sure have to have the pure stuff
separate for 'em. Bueno?"

So, we got things all taken care of, put the different
calibres into labeled piles and went to supper. I didn't get
to sleep real early that night. Finally, though, I figured out
that, hell, we were just setting up drinks for the house. And
with that comforting thought in mind, I drifted off.

It was an evening sale, so the next day we gave our stock
a final going over. J. D. disappeared along in the afternoon,
but a while before the sale was to start, he showed up with
a pickup full of pop, ice, buckets, a metal watering trough
and six or seven kids. Under his supervision, they put the
high-power pop and ice in the watering tank and a couple
of buckets of ice and low-voltage mix up on the auction-
eer's and clerk's stand. Then a couple of buckets of plain
Coke well off to one side. The seats were beginning to fill, so
he called up his helpers for their final instructions. As the
old boy from down the Yellowstone and I listened, fascinated, he told them, "Remember now. Keep movin'
through 'em. Offer 'em a Coke whether they look thirsty or
not. That big bucket yonder is jest for you fellers. I catch
you drinkin' out o' any other, an' I'll tan your hides an' nail
'em to the wall. Savvy?"

"Yes, *Sir*" was their respectful answer. He sure had them
buffaloed. J. D. was a fair man. Just before the sale
commenced, he told us, "I got more stuff to sell than you
two, so we'll sell some of mine first. Then, time our sody
pop gets into gear, we'll take turns."

That's what we did. Though things started a little slow,
pretty quick the tempo began to pick up considerably.
Matter of fact, time J. D.'s corps had made several trips to
the trough, even the ring men were having trouble keeping
up with the bidding. Toward the end of the sale, a man
could almost have said it got feverish. Oh, there were a few

arguments in the stands as to just who the hell had the last bid, but everything got settled amicably. Evidently, the booze J. D.'d bought was a friendly tipple, though it sure wouldn't have been the choice of a connoisseur of whiskey. At least, everybody stayed happy.

It was worth the price of a ticket just to watch the old scalawag in action. My partner and I plumb hated to leave the ring for our horses when the old devil was showing one of his. Talk about sanctimonious—a man would have sworn he was a self-ordained lay preacher—and when he was enumerating an animal's good points, his voice was full, rounded and utterly convincing. Resonant, sort of like a man talking down a pickle barrel.

So the sale wore on, maybe a little noisy but damn sure not slow. The prices were not sky-high by any means. Just good solid prices for damn good horses, and when, in what spare time I had, I figured my average, I found my stock was doing a good deal better than I'd hoped for. The boy from the Yellowstone was doing about the same. J. D. though, seemed to be outselling us both.

The pop ran out a little before the sale did, but the momentum it had generated kept on to the end. Then the stands emptied except for a few individuals who were snoring peacefully. The ring men were a little thunder-struck at the way the sale had gone, but J. D. assured them, "Hell, best string of horses you ever he'ped sell, that's why."

Gibby, of course, claimed full responsibility for the success. Naturally, he'd had to keep his throat moist so he could perform well and consequently had dipped into his low calibre pop frequently. We damn sure didn't argue with him.

A man who'd looked at my horses about a month before walked up, packing a happy grin. "Well, Spike, I bought those two horses I thought you were holding a little high that day at the ranch. Should 'a' got 'em then. Cost me more here. I like 'em, though."

He frowned slightly. "Funny. Damn if I don't believe I raised my own bid there a time or two. By God, though, I sure didn't want to lose 'em."

Finally, the crowd drifted off, and J. D. turned to his helpers. "You boys corral th' empties an' put 'em in the trough. A couple of you wake them fellers in the stands, too, an' he'p 'em down."

When they'd finished their chores, he said, "You did a good job, so here's two dollars apiece. An' take the rest of that pop by the door. You didn't have time to get much of it drunk up. Take care an' thanks." And they departed happily.

When they were gone, he turned to us. "What'd I tell you? Hell, I *know* about horse sales. I been there. We did fine, an' so, by God, did th' buyers. Now, happens I saved one o' them bottles of whiskey back up in my room. What say we go sample 'er. It's been dry work leadin' all them ponies." So we did.

During the '50s and early '60s when I was running quite a string of horses, I was involved in several more sales but never where the ramifications were the same. Never, too, where my stock sold any better. My conscience has never bothered me a hell of a lot about that first sale either. Particularly, since my buyers were tickled to death with what they got. Hell, twenty years later they were still bragging about how well their colts turned out. Just last fall one of them called to ask if I had a couple more I'd sell him.

J. D. was right—we sold horses that were sure worth what we got for them. On top of that it was one of the damndest, funniest deals I ever ran into!

===11===
BEDS OF PAIN

IT SEEMS TO ME that I've spent more than my share of
time in doctors' offices and hospitals—not because I was
sick but more on account of the fact that I had quite a
penchant for taking second money when a horse and I had
a bad falling out. It got to be sort of monotonous. Even Doc
Baskett in Big Timber sewed me up so often that he finally
quit asking me what the hell had happened. He knew. As I
got older, slower and more brittle, my wrecks couldn't be
patched in an office, so a hospital was the next step. I got
pretty well acquainted with a number of them: in Big
Timber, Billings and Rochester, Minnesota. I never really
liked any of them, though some were better than others—
depended on the nurses, mostly.

At that, though a hospital is not a particularly shiny
home away from home, a man's attitude and outlook can
do a lot to chirk up the sojourn. Matter of fact, looking
back I had some fun during most of my stays. At least, I
tried to make the best of a sorry situation, and, by God, it
seemed to work. Of course, I never put in much time in
what you would call, I guess, medical sections, and when I
did, it was pretty grim: poor guys who didn't know what
was wrong with them; doctors like owls and just about as
uncommunicative; the tubes and paraphernalia they made
their patients wear, swallow, or just plain endure. Gave me
the willies.

I spent most of my time in orthopedic sections, and they

are different. Everybody knows for sure what is wrong with him. Everybody hurts, but knows *why,* so we were all even. I've always been damned if I'd stay in a private room where I could stare gloomily at the ceiling and reflect, all by myself, on what was wrong with me. Or even in a double room because God only knows what sort of a roommate you'll draw.

I was in one only once and drew a sick doctor for a partner: the whiningest, most self-centered, loosest-boweled individual I ever had the bad luck to run into. He lay complaining all the time or moaning in what he called the "fetal position." Except when he had to be helped to the bathroom and back by a long-suffering nurse. Which was about every fifteen minutes. Whatever the hell was wrong with him sure didn't show, so I finally told her when she brought in our trays while he was enthroned, "Why don't you give his to him in there? The bastard won't be done with his food out here before he has to go back in. Besides, for a change I'd like to be able to eat without having to listen to him whine." She wouldn't do it, but I got moved to a ward the next day, thanks to her. She told me, too, that the man was a shrink.

In a ward, particularly an orthopedic one, there is always at least one person who is cheerful and likes to talk; and talk takes your mind off your troubles, so there is something going on most of the time. The guys all are in more or less of a similar bind, and most of them have something of a sense of humor, though it might be a little macabre. Either that or I've just been lucky in the wardmates I drew.

My first real hospital experience was when I went back to the Mayo Clinic to get a hip worked on. An old pony had jimmied it badly. I know now how a critter feels after he's been worked with the herd in a set of pens. How they can handle all those people and find out exactly what's wrong with each of them beats me. But they sure do. It's a humbling experience. I got there feeling pretty sorry for myself, but after seeing all those poor devils and the shape they were in, I damn near just went on home. I didn't have anything wrong; in comparison, that is. If you think a man can get badly stove up by a horse, you ought to see what a corn picker can do!

They fixed me up in St. Mary's Hospital and put me in a ward. I was a little woozy for a day, but about that time another Montanan was put in the room. Lucky, too, for if those nurses had run into just one of our breed all alone, they'd never have believed it. As it was, it took a little time before they decided we were for real. He was from Butte, and denizens of that camp of his vintage are about on a par with Melvilleites. We two had a lot of fun, so did the other guys and the staff. I think. One girl did tell me though, "Spike, I wish you'd stop your swearing."

I explained, "I don't swear. I just cuss a little."

"Maybe so," was her answer, "but I'll never see a washcloth again without thinking of it as 'damn wash-cloth'."

The head honcho of the section, a big, redheaded woman with a lot of bark on, told Maurice and me, "You two are troublesome bastards, but cheerful. It seems to be rubbing off on the other guys. And on the nurses, too." But we sure got blamed if anything went wrong, nevertheless.

Like when they brought Maurice back from the operating room and put him to bed. They either forgot to put the side bars up or had misjudged how much dope it took to cold cock a guy from Butte. Anyhow, pretty quick he began to stir and mumble, got hold of his monkey bar and began to work his legs over towards the side of his bed. There wasn't a man in the room that could get around afoot, so we all yelled at him, but he was dead to the world. We pushed our buttons with the usual result. Nothing. So we held a hurried confab, and somebody came up with the idea of each of us getting whatever stainless steel utensils that were in our tables and empty. Then, all together we threw them at the open door. The floor was made of those little pieces of marble or something in concrete, and the noise was magnificent. And effective.

Just as his legs started to slide over the side, a nurse appeared at the door. "Spike!" she snarled, then, "My God," and she caught him before he went any further. A couple more girls hurried in, and they got him back where he belonged. This time they fenced him in, and when the excitement was over, the girl told us, "Thanks, guys." She needn't have. We didn't want him to get hurt.

Actually, I don't know for sure what they had to patch up on Maurice. He told us a twenty-foot, six-by-six mine timber had tipped over and hit his head, but, hell, you can't hurt anyone from Butte by hitting him on the head. I believe it was his back that they'd operated on, and he'd have really fixed things if he'd gotten out of bed.

The Mayo Clinic is real popular with South Americans to judge from the number of them that were patients, so we had a Colombian in our ward. Bill Abadi was his name. A nice guy, and I got a chance to practice my Spanish. I forget whether his hip or knee had been worked over, but he had a sort of roller skate on one foot and a piece of plywood on top of the sheet on the lower third of his bed. He had instructions to roll his foot up and down on it, and he sure did. It made a hell of a racket. Trouble was, he got to doing it in his sleep during the night, and we'd have to wake him up so's we could get some sleep ourselves. Sort of a "ride and tie" deal—we'd sleep a while with him awake, then he'd sleep with us all wide-eyed.

Then there was an old fellow we called "Uncle Ed." He was deaf as a post—the only one in the ward who could sleep all night—but he was a nice person. He never said anything; why should he when he couldn't hear anything and just sat and watched. And smiled. We liked him and his wife. She'd come every day, sit by him, hold his hand

and smile, too. I got a Christmas card from them every year until he died. A fine old couple.

We had a man from Washington state, too. He was in a cast from his neck to his knees. Made us all feel how lucky we were and a little ashamed, he was so good-natured and uncomplaining. Why, when the nurses bunched up and turned him over, which was about four times a day, it sounded as though they were rolling a beef over. Dwight was all right. Gutty as a barrel of angleworms.

We had an empty bed for a while, and one evening they brought a new man for it. We all tried, except Uncle Ed that is, to draw him into some sort of conversation, but he was sour. He only had a broken arm anyhow, which really didn't qualify him for our outfit. He didn't stay long. Possibly because when a nurse's aide wandered in looking for someone she was supposed to give an enema to, we pointed her at him. Of course, he raised quite a fuss but wasn't in shape to put up much of a fight. We plumb convinced her that he was her man, and he got the works! He left the next day, still mad. The girl came in later and gave us hell because it'd been a man down the hall she'd been supposed to favor with her attentions. We all enjoyed the deal. Besides, we figured nobody had any business in a hospital without an enema, anyhow. They sure hadn't passed any of us up when we came in.

That afternoon they brought in another man for the empty bed. He was in bad shape, judging from the way they walled him in with sandbags from waist to head. We all took a lot of interest in the proceedings, sort of like a new kid on the block or a strange dog on the ranch. But we kept our voices down so we wouldn't bother the guy. God knows we had plenty of time to get acquainted with him when he felt better. So, keeping it down, we got on the subject of our nurses and which of them we liked best. They were all nice girls, but we finally agreed on one. Maurice, because she was so pretty—a natural reaction for a Butte native; Bill,

for her "felicidad"; while I appreciated her friendly disposition. Uncle Ed had no idea what we were talking about
and consequently, no opinion. Finally, Dwight spoke up
from the depths of his body cast, "She's so nice and
sympathetic."

"Sympathetic?" came a hollow voice from the newly
occupied bed, "By God, send her over. I need sympathy
bad."

We all chuckled and introduced ourselves. His name was
Brown, a farmer, feeder and buyer from Indiana. It
surprises me that I can't come up with his first name, for he
was a corker. I think it was Jim, and his wife who came in to
see him was Naomi. They were a great pair. He'd had a
bunch of vertebrae fused, and in spite of the fact that he
had to lie absolutely still, his mind damn sure didn't. Why
in no time at all his reputation with the staff was as bad as
Maurice's and mine. On top of that, he was a mule man, so
he and I went round and round in our spare time apropos
the comparative merits of horses and jackasses. According
to the others as unbiased listeners, we broke about even.
He couldn't do anything himself, but he was a persuasive
cuss and talked other people into the damndest things. The
jackhammer deal, for instance.

One day for some reason the hospital got a crew to start
breaking up the concrete below our window. It was June,
the windows were open, so the racket was bad. Seemed to
bother the nurses 'way more than it did us (by that time we
were all pretty used to taking things as they came), but
when he figured the girls had all gotten sufficiently edgy,
Jim made a casual suggestion. It was pounced on, the
window screen was taken off, and, sure enough, the
jackhammer crew was hard at work right below it. So with
the mule man ramrodding everything from flat on his back
and a few suggestions from the rest of us, the girls filled the
biggest wastepaper basket in the ward with water, hurried
it to the window and, calculating the trajectory, careful

not to get any water on the wall outside which would tip the location of the bombardment, upended it over the sill. I can still see the bevy of starched white seats bent over the ledge. One girl damn near went with the water, she was so intent on her aim.

It takes a little time for water to fall three stories so there was a moment of hushed expectation. Then a satisfyingly wet explosion, the jackhammer quit, and a heartfelt burst of profanity rose from below. There was a burst of giggles, and I got wide grins when I said, "And you accused *me* of swearing." None of them looked very shocked either, even though the language was pretty juicy.

I believe everything would have been fine except that one girl, typically feminine, couldn't resist peeking out to see what had happened. She was spotted, and the fulmination outside soared to new heights. They bayed, and none of it was complimentary. It even embarrassed us, the things they bellowed at her. She jerked back, the screen was replaced and at Jim's order the basket was rushed into the johnny, dried with toilet paper and put where it belonged. The nurses poured out our door like, as the man says, "A giggle of Girl Scouts," and scattered through the section.

Silence for the first time since morning. Then purposeful steps came briskly down the hall, and in came the head of our section. Miss Van something-or-other, the big, red-headed woman. "Who threw that water?" she asked. We guys knew her pretty well by then, and she didn't sound really on the fight. More like she was curious, and I got the feeling she hadn't enjoyed the hammer either.

We all claimed to have been asleep. "Oh, sure," she agreed, "with all that damned noise. Come on, who was it?"

Silly question to ask a roomful of people, none of whom could possibly get out of bed, but Dwight answered. "I cannot tell a lie. It was me."

She sniffed. "You know what I mean. What nurse was it?"

Jim told her, hell, he couldn't see anything but the ceiling. Maurice hadn't gotten a good look at whoever it was, but Bill came up with the clincher. It had been a nurse, true enough, but "una extranjera." Uncle Ed, whose bright eyes had been following everything, saw Bill shrug, so he shrugged too, and Dwight burst out, "By God, that's so. I never saw her before in my life."

I jumped in with, "That's a dirty damn trick. To come from another floor and try to get us and our nurses in trouble. Hope you catch 'em."

Her gaze swept around the room, she shook her head and walked out. But just before she got her back turned completely, a trace of a smile showed. It was the last we ever heard of the affair, so we suspected she'd taken a cue from us and had covered for her girls.

That head nurse was all right. When after about a month I was up on crutches, Barbara came back for me. We weren't leaving until morning, so when I left the hospital, I asked if I could come back that evening and have a party in the ward. Red looked suspicious, "Don't bring any booze."

"Nope, but how about beer?"

She allowed that, so Barbara and I got a case of beer downtown. Then we got our taxi driver to bring it up on the freight elevator and carry it into the ward. Just to be on the safe side, I had a bottle put in the refrig for Red. With her name on it, for she was off somewhere.

Maurice had left a couple of days earlier, so there were only the five of us and Barbara. She didn't drink much beer. Just sort of kept an eye on us. Bill was on crutches by now but had to go back to bed before he could have any so we were sure he wouldn't have a wreck. The thing was I'd gotten beer all right, but it was called "Stite." Sort of like ale and stouter than all hell. I'd gone to some trouble finding it, too. So Jim and Dwight got cranked up to where they could drink without drowning themselves, a round was uncapped, and we settled down for a last talk. All but Uncle Ed. He just drank and smiled.

It was a good party. Though time we'd worked through the Stite, its effects were showing. Bill had progressed from English to a mix of that and Spanish, to straight Spanish and was singing Colombian songs. Dwight gradually got sadder and sadder. Then he started to cry because the old gang was breaking up. I threatened to put itching powder down the inside of his cast if he didn't quit. That tickled him—figuratively, I mean—and he got to giggling. Uncle Ed, brighteyed as a chickadee, downed bottle after bottle, though he did spill some of the last on his bed. His wife even had one when she came in. He wasn't smiling any more. Instead, he had a grin that threatened his ears and said the only thing I'd ever heard him say. Told me he was going to miss me. Jim was going to come see us at the ranch. He'd bring his favorite mule, and we'd have a contest. I don't remember what sort, but it'd damn sure be a horse-mule deal.

Finally, we ran out of Stite. The nurses were worrying some. At least, they kept appearing in the door with anxious expressions—though I'm not sure whether it was about us or because they expected Red back—and Barbara said it was time to go, anyhow. So I said goodbye and good luck to each of the guys, shook hands and we left. I've never seen any of them since.

Funny. A man spends upwards of a month with a bunch of men like that, and you sort of get to be a family. Why, hell, when I got home, be damned if I didn't miss them and the ward for a while. We wrote to one another, but that sort of died off, too. I'll never forget them though. They were a good outfit.

I spent some time in a ward at St. Vincent's in Billings. I was there only about a week—horse trouble again—but I got a boot out of some of the things that happened. One of us was a great big, old man. He, too, was named Ed. He was there to die, I guess, for he was in damn bad shape. His mind wandered, he didn't know where he was most of the time, but now and then he'd fight his way up to reality. A nice, tough old guy. Another was an individual who knew everything about everything and wasn't a bit loath to tell all about it. When we got used to his compulsive talking and learned to say "Sure enough" whenever he'd stop to get his breath, we did fine. He was a good-hearted booger, and at least the ward was never depressingly quiet. Then there was a real nice young fellow from West Virginia. A truck driver who'd gotten crippled in some sort of wreck. The fourth was a rancher from over in the Missouri Breaks. He was a typical old-time cowman and was in to get an eye taken out.

Anyhow, one afternoon the talker was giving the truck driver a verbal tour of the West and suddenly asked, "You ever been to Thermop'?"

Ed must have been having a lucid spell because he answered stoutly, "Sure. Lots of times."

The raconteur with some annoyance snapped, "I wasn't talking to you."

"Why the hell'd you ask me then?" the old fellow came back belligerently. It stopped Windy cold.

When Barbara would come to see me, she'd always go visit with each of the others. Ed hadn't the slightest who she was or where he was for that matter, but it always pleased him. He'd perk up and talk to beat the band to her, though mostly it was about things long past and places far away. Being the lovely person she is, Barbara made a point of their conversations. Windy, too, waited for her to come over to his bed like a cat at a mouse hole and bloomed like a rose, or an encyclopedia, as they chatted. In no time at all, she was "Barbara," and you'd have thought they'd gone through school together. West Virginia was a different kettle of fish. Shy, quiet, and never failing to address her as "Ma'am" or "Mrs. Van Cleve." He enjoyed the visits.

"Spike," he said after she'd left one evening, "long time ago my pappy tole me all women was ladies, an' never to forget it. Don't reckon I ever seen a woman as much lady as Mrs. Van Cleve. Nor never will."

She and the rancher from the Missouri Breaks were thick as thieves. They talked cows, horses, prices, grass and had a fine time doing it. Not surprising, since Barbara savvies ranching as well or better than I do. They finally operated and brought him up to the ward to come to. Probably the recovery room was full or something. While he was dead to the world, in walked Bob Langston to pay me a visit. He and I sort of work as catalysts for one another, and it wasn't long until our end of the ward was booming. Windy was pretty near busting a gusset to get into things, so Bob finally went over to talk to him, too. While they were at it hot and heavy, in came a pair of nice women. They had come to see the rancher. Seems they were from over in his country and being in Billings, had thought they'd say howdy. It was real kind of them. Made what I did next, pretty bad. One of them asked if their friend had had any spiritual comfort while he had been there. I disclaimed any knowledge as to that but told them that the man over talking by that bed was, I believed, some sort of a lay preacher, and maybe he could enlighten them. So they hurried over to Bob.

I knew my man! Inside of a minute, his face mirrored pure sanctimoniousness. I couldn't hear what he said, but I could catch the tone. Unctuous, like butter wouldn't melt in his mouth. I'd have sworn he was an old-time undertaker. The ladies seemed eminently satisfied, however, and soon left. Happy as larks. When they were safely gone, Bob turned to me, "Damn you. Give a man a little time before you spring something like that on him."

I grinned, and Windy, who'd been listening to their conversation open-mouthed, asked hesitantly, "Are you sure enough a preacher?"

It was too much for Bob. "Well, yes," he answered without the least hesitation, "but self-ordained, sort of."

"I be damned," was all his questioner could say. Matter of fact, he didn't let his full weight down, when it came to talking, for the rest of the day.

Terrible thing for me to have done, and I felt ashamed of misleading two nice women, but I chirked up the next day. After the payoff.

The old rancher was awake and feeling pretty good when in the door stalked a tall, somber figure carrying a book that would have made a Monkey-Ward catalog look like a pocket notebook. Standing at the foot of the old fellow's bed, he opened the tome and began to read. If I'd shut my eyes, I'd have sworn it was Bob yesterday. Only it was louder. He boomed on for maybe five minutes, shut the book with a crash that made us all jump, spun on his heel with his coattails snapping and left. Not another word. The rancher whom he'd been praying for or at, his one eye completely surprised, looked at the rest of us. "Goddam. Who d'you figure that was?"

My girls all rodeoed with me when they were growing up, so some of the contestants who'd been to a real good roping at Geraldine decided they'd pay Carol a visit on their way home. She was in the hospital at Big Timber with a badly broken thigh, and the boys figured on cheering her up. They did. I wasn't there, but, believe me, I heard about it. From several sources.

Well, they'd won some money, drunk some whiskey and had bought a huge bouquet of flowers. They trooped in, spurs jingling, and asked the nurse, "Where's ol' Buckshot at?" She was new to the country and didn't know who they were talking about, so some slight difference of opinion arose. Voices got a little loud. Then it was decided, "By God, we'll find her our ownselves!"

Down the corridor they went calling for Buckshot, a distraught nurse fluttering at their heels. Fortunately

there wasn't anybody real sick in the place, though visitors
to other patients poked their heads out of doors trying to
find out what was going on. One look at the outfit was
aplenty, and they popped back again. Finally, the boys
found Buck, who'd been listening happily, and crowded
into the room. They sized up her weights, pulleys and the
rest of the paraphernalia, and things calmed down. Two,
who were pretty well the worse for wear, were sat down in
chairs well away from the bed and told in no uncertain
terms to stay there. The rest took off their hats, hunkered
down and went to visiting. I understand that periodically
over the next hour or so there were roars of laughter, and
whoops of glee or derision from the room. Each time, the
nurse would hurry down and nervously peek in—to receive
an enthusiastic greeting. Finally, the stories ran down, the
cattle at Geraldine and other shows had been roped until
they were played out. Even the "D'you remembers." So,
waking up the guys in the chairs, everybody filed out with
last good wishes.

"Thanks a lot for coming, guys. It was fun, and the
flowers are lovely," Carol called after them, tears in her
eyes. Not just from laughter either. She meant it!

Our local hospital is a pretty broadminded outfit, but I
guess it got a little strained that evening. I gather, too, that
the one nurse stayed a little jumpy until Buck, at long last,
went home.

Speaking of hospitals, one last story comes to mind that
is just too good to keep quiet about. It happened at the
Dude Ranchers' Convention in Billings several years ago,
and it was just my luck to learn of it only after the smoke
had cleared. Had I known about it, I sure would have tried
to compound it. It seems that a good friend of mine, I'll call
him Dick, an attorney and associate DRA member, was
the main figure in this jackpot. He's a good guy and has had
himself a fine time at our conventions over the past years.

Seems he didn't feel too well Friday morning after the

Thursday night festivities at the Northern—he's strong on social hours—and as the day wore on, being a lawyer, he convinced himself that he sure had something wrong with him. So in the afternoon, he decided he'd forego the horse discussion that was scheduled and go to the clinic for a checkup. He'd suddenly realized he hadn't had one for pretty near six months. Well, the closer to the outfit he got, the worse he felt; by God, maybe he had flu, too, come to think of it—he'd better get a shot for good luck. So it was quite a letdown to find that the doctor wasn't in. I forgot to mention that the guy has eyes like a seal pup and by that time probably looked pretty forlorn, so the nurse very solicitously suggested that he go on up to the emergency room at the hospital, and they'd take care of him. She wasn't just woofin' either!

As he drove over, I sort of imagine he whiled away the time studying over his symptoms, and when he walked in, he must have been as wet-eyed and humped up as a toad in a hailstorm. Anyhow, the man in charge asked what was the trouble, and Dick replied, "I just don't feel good and thought maybe I better get a flu shot."

The doc sized up those big, bloodshot eyes and woebegone stance, must have figured he had a red hot prospect and ordered, "Strip. I'll be right back."

There were three nurses in the room, so Dick, time he got down to cases, decided he'd cheat a little and kept his drawers on. One of the nurses reminded him the doctor'd said strip, so he sort of fiddled around with the waistband until the three finally lined up and headed for him. The doctor saved him by coming in just then, so Dick used him as cover as he got out of his drawers and into one of those runty, backless hospital riggings. "Get up on this cart," said the doc.

Gingerly, clutching his gown tight, he did, and sighed in relief as they put a sheet over him. But his relief was short-lived. "Take him to Intensive Care" was the order.

The trip to his room was a long one! There he was tucked in bed, the doctor gave a few crisp instructions, and in short order old Dick sprouted all sorts of things—an IV from his arm, an oxygen tube from his nose and a maze of wires from different parts of his anatomy to the big electrocardiograph machine they wheeled in. Hell, they should have known his heart was in good shape or he'd have died right then! Still, I can't help wishing that as long as they were doing all these things, they'd given him an enema, too.

After a while, they shut off the machine, the doctor shook his head ominously over what it had recorded, turned to Dick and said, "Better call your family."

Well, the poor devil could imagine what a call would do to his family under the prevailing circumstances, so he quavered, "Good God, no! I don't want to scare them, too. But there's a friend in town I am supposed to meet this evening, and I'd like to call him so somebody'd know where I am."

That's why the phone in our room rang. I was at a meeting, but Barbara answered it, identified herself, and a hollow voice said, "This is Dick, and I'm in the hospital." Then the voice cracked as it added, "In Intensive Care."

"Oh, no!" said Barbara. "What's wrong?"

There was a moment of silence; then, in a shaken tone, "By God, I don't know."

The two talked a few minutes; he promising to keep us informed; she to call his wife when something definite was decided; and just as he hung up, his own doctor came hurrying in.

"What's all this, Dick?" he asked and then examined the record the machine had made. Then, "I've got to run another one," so old Dick was once again hooked up. He lay quaking under the sheet until his doctor shut the outfit off and scowled over the new record. "Godamighty," he suddenly exclaimed, and left the room abruptly.

"My God," thought the patient. "What now?"

Providentially, before his imagination could run much further amok, in came the doctor again. "I don't understand what happened, but there's nothing wrong with you," he said. "Get out of here." Then, surveying the pallid face, he grinned and remarked, "I'd say that if there was anything wrong with your heart, there's no way it could have helped showing up this last test. The hospital's lucky you didn't die though, at that."

I saw Dick that evening, and he looked a little shad gutted even then. He complained bitterly, too, about how long it had taken to get his wallet back after he'd gotten dressed. Acted like the sooner he'd left the hospital, the better! Since the deal was over by then, and I hadn't had a hand in it, about all I could do was to inquire as to whether, as long as he was at the hospital, he had gotten his flu shot. His response sort of eased my sadness at missing out on the proceedings!

====12====
SNOOSE

ALONG WITH LIVESTOCK there's another thing I've been around all my life: snuff. Due to the proximity of the "Settlement" and the many newcomers that worked for us, I accepted it as a fact of life. For Norwegians, that is. Most everybody else smoked: Durham, PA, Velvet or maybe Tuxedo. I'm surprised how many people use snuff today. It used to be considered sort of nasty, but not anymore. Now there are sterling lids for snuff boxes, new exotic varieties, women afficionados, you name it.

When I was a kid, snuff meant "Copenhagen" or "Snoose." It was years later that I found out that snoose wasn't spelled like it sounded but actually is *Snus*, the Norsk word for snuff. Now, however, with all due respect to Walt Garrison and the U.S. Tobacco Company, snoose has deteriorated into such stuff as Skoal, Happy Days and other triumphs of insipidity. The former, Skoal, would make an old-timey Norwegian quit booze or worse. Add a little sweat to its wintergreen flavor, and a man'd swear he was in a high school locker room. Happy Days is too foul for a true snuff man to contemplate—raspberry flavor! The mind and taste buds boggle!

Those of us who *know* are Copenhagen men: the one, the original, the only snoose. U.S. Copenhagen, by God, not Canadian. The latter is different somehow. Not really true Copenhagen.

Of course, I know that Copenhagen is a far cry from

154

what was originally called snuff. Matter of fact, it's a far cry from what the Southdowns in Appalachia use. They "dip" snuff, not chew it. It is a fine powder, flavored and suited to the original raison d'être of snuff: snuffing.

Actually, I believe the latter type, and habit, were pioneered by the Incas—though I sort of wish it had started in Scandinavia. At least, Pizarro mentioned it as far back as 1532. Then, along about 1560, the French ambassador to Portugal, Jean Nicot, imported tobacco seeds to France. The word *nicotine* was a direct result. At any rate, he talked Catherine dé Medici into trying snuff—possibly a deep-down, keystone contributing reason for the dé Medicis' and the Borgias' nasty reputations as to poison a little later on.

Snuff was popular in the courts of Europe during the 18th century. Frederick the Great, Queen Anne and Catherine the Great were all addicts. George III's queen was known as Snuffy Charlotte—which may, or may not, have helped the American Revolution. It's even said that Napoleon was so dependent on tobacco that all his portraits show him reaching for the snuff box inside his jacket.

Maybe so, but at the height of its popularity, it's a dead cinch from what I've seen and read, that the proffering of the snuffbox and the accepting of a pinch of its contents were graceful, almost ballet-like gestures. Courteous. It surely was among the old-time Norsks. The snoose can was produced, the lid was tapped gently with the knuckles of the left hand, then rapped a time or two against the left elbow. The lid was removed in the left hand, and the opened can extended with the polite query, "Skal du ha snus?" Invariably, a substantial pinch was taken between thumb and forefinger and tucked in the lower lip with a mannerly, "Taks skal du ha."

Sadly, it is no longer thus. Now it is, "Gimme a chew," or an uncouth offer, "Ya wanta chaw?" Just about on a par with these "cowboys" I see dancing with their hats on.

Hell, when I was growing up, most girls wouldn't dream of dancing with a man so impolite as to wear his hat while he was doing it. Besides, he was damn liable to get knocked clean out from underneath it by some guy who had at least the rudiments of decency. I guess though, hell, if the modern girls don't mind, why should I?

As I mentioned, the Norsks I knew put their chews in their lower lips, with the result that in time they all had lips resembling those of old work horses. I guess that's why I always put snoose in my cheek; I wasn't handsome to begin with and could ill afford a thick, sagging lower lip. Naturally, I tried it as a kid, but it paralyzed my jaw for about half a day and spooked me. So badly, in fact, I didn't tackle it again until I went to prep school in Massachusetts, better than fifty years ago. I've used it ever since.

Funny how I started. The school banned the use of tobacco, so, of course, we all tried to circumvent the rule. The other guys smoked, eventually got caught and paid the piper. However, there was a small Norsk settlement nearby. I remember that two of them, the Satre brothers, were ski jumpers. So taking advantage of a Sunday spent in Great Barrington, I discovered that one little store there sold Copenhagen. I had it made—after a couple of go-arounds with paralysis of the jaw. One box would last a week, and though we had a sneaky, devious housemaster, I resorted to caching my can inside the base of my desk lamp, which thwarted his attempts to apprehend me. He made a lot, too, for I spit out my window all the time, and the resultant dead grass roused his suspicions. To no avail or I'd probably have come home to Montana sooner than I did.

Barbara, after years of stalwart dedication, has weaned me from most of my vices—I don't call 'em that, they are hobbies is all—except snoose. She quit trying years ago, but I've had a couple of abortive tries at quitting by myself, and I guess I got pretty cranky when I did. But I quit quitting when I announced at the dinner table years ago, "I think I'll swear off Copenhagen."

The heartfelt answer I got from Barbara and all the kids was, "Oh, *please* don't!"

Actually, I don't know whether I could quit, but since I don't figure on it, by request, why the hell should I.

It's powerful stuff. I had a herder tell me of pinning down a rattler, prying his jaws open to where he could pour a good shot of snoose into him, and within ten minutes the poor damn snake was dead as a doornail. It's especially virulent when combined with booze. I learned long ago never to use it when I had drink taken. Lord, if I've had a few and take a chew, in no time I have trouble staying upright, even with a death grip on the bar.

I've heard it called by non-Norsks "Skandihovian condition powders." Maybe they've got something, for I've seen herders, particularly when they come out of the mountains with their bands and are celebrating their fall drunk, pour a shot of whiskey into their snuff can, mix it well and take a big chew. Doesn't seem to bother them *then*, but I've noticed that damn often it is a gilt-edged guarantee to a howling drunk, first-class hangover and usually a thorough stomping to boot. Sometimes they'll put in a little lemon juice, red pepper or baking soda to try and perk up snoose that has gotten old and dry. But a true afficionado will buy, beg or steal fresh snuff when his current supply gets old and tasteless. I've always been a conservative, middle-of-the-road user, so I don't know whether the doctoring really works.

Then there's the story of how a ship was wrecked, and the only survivors were a good-looking, lusty girl and a Norsk sailor. They made it to a little, uninhabited island; he built a shelter for her on one side and one for himself on the other. Things rocked along for some six or eight months, but it got a little too much for her. So she went over to his side and accosted him with, "Sigurd, I have something I know you want and . . .," but before she could finish he leaped to his feet and said ecstatically, "Oh, Yeesus, lady. You got snoose!"

I've slipped it to a dude now and then, and the results have been gratifying. Especially, if they've snuffed it: explosive sneezing, watery eyes and noses resembling a kid's red toy balloon for hours, sometimes even a day or two. Mostly, though, all I need to say when they notice the can in my pocket is that I'm afraid they better not mess with it. Then it's a dead cinch they'll insist, so I feel my skirts are pretty clean.

I think the funniest thing I ever saw as a result of snoose was when I was back at the Mayo Clinic. An old pony had turned over on me on a mountain, and I couldn't get rid of him so they were cobbling me back together. I was all patched up in St. Mary's Hospital and had talked a ward mate who could walk into getting my Copenhagen out of my shirt in the locker. So I had a big chew and was as happy as a skunk in a chicken house until I had to spit and realized I couldn't get out of bed. Finally in desperation, and because it was handy, I used my urinal as a spittoon and leaned back happily.

I guess it was about an hour later that I got rid of my chew, same place, had a glass of water and used the urinal for what it was meant to be used for. A few minutes later our little nurse's aide bustled in, took the thing and went into the bathroom to empty it. The result was exciting. She popped out the door, pointed the urinal at me and asked breathlessly, "Is this yours?" just as though she hadn't taken it from my table a minute before.

"Yup," I answered. "Why?"

No answer, but with a horrified expression she hightailed it out the door. Shortly she was back with a veritable bevy of doctors. Their expressions tickled me. A couple looked genuinely concerned; the oldest, evidently the head honcho, showed a little interest in an Olympian fashion. And the young ones gave me the impression of happiness—here was something new, a glorified medical crossword puzzle, sort of. Into the johnny they trooped, and everybody in the ward could hear sounds of argument. Then out they filed, the last, most junior of the group, packing a sample of what was in the toilet bowl. "That yours?" asked the jefe.

"Used to be," said I. "Why?"

That got 'em. Nobody had an answer. So, because I didn't know how long I could keep from laughing, which would hurt my crippled hip sure as hell, I told them about the snoose.

Silence. Then there was a snicker from the tail-end Charlie who had the sample. A glare and a "humpf" from the leader, and out the door they went to the accompaniment of a roar of glee from the ward. I had to join in, hurt be damned.

I get a boot out of the sterling silver snuff lids so prevalent now, though Barbara, bless her, has asked if I'd like one. I said, "Hell, no." Unless I remembered to remember, I'd probably throw it away with the empty can. We didn't during World War II, by God. The country must have been hard up for metal then, for all the snoose cans had cardboard lids instead of the usual tin ones. They were a sorry proposition. They mashed easy and fit so sloppily that, too damn often, a man'd find himself with a pocketful of loose snuff when he reached for a chew. Anybody who'd saved or rustled up a tin lid was in the money. I've seen them sell, back then, for a dollar apiece. Why a man could keep in snoose by working over the trash barrel and peddling the lids he found. A lot of people did. In those days, Copenhagen sold for a dime a can, eighty cents a roll. Now it's pretty nearly eight dollars a roll. Tough times, tough times, when snuff costs better than six bits a can. But we keep on buying it—probably would at a dollar.

Besides the sagging lip, you can usually tell a snoose man from the worn circle in the front of his shirt pocket. That's another reason women are against it; wives, that is. Hell, I've had shirts that were just nicely broken in, and Barbara has made me get rid of them because a circle of cloth had fallen out of the front of a pocket. Even when I promised to swap pockets, so's in time they'd match up.

I've made lifelong friends with snoose. I had a string of guests on a ride to the head of the south fork of Big Timber Creek. When we got to the upper meadows, here was Trygve Sandsdal with one of Grosfield's bands. He seemed nervous and shaky as he talked to us. I savvied when he turned to me and asked pleadingly, "Have jou snoose, Spike?"

I said sure and gave him a chew from a can I'd opened that morning. Seemed Cris Vik, the camptender, was overdue, and Tryg had run out of Copenhagen the day before. I left him my can when we rode off. From then on out, whenever I met him in town, I had a free drunk coming if I wanted!

I've made an enemy or two, too. All cans carry the date and month they were packaged, and all good snuff men buy it as fresh as possible. Nowadays, the year is shown, too. In the old days, once in a while when snuff got too old and stale, I'd save a virginal can until the date rolled around, and then slip it to some unsuspecting guy. The further from town he was, the better the joke. Trouble was, some of them took a dim view of it and be damned if I can blame them. Imagine opening a nice, new can of snoose and finding it was anyhow a year old—especially when a man needed it bad.

I've known men out of snoose to do some funny things. More than once, a good, responsible herder, snooseless, has just up and quit his band and walked to town for snuff. Sometimes twenty miles or more. I've run out a time or two myself, but my thought is that it isn't the running out that's so damn bad. It's *knowing* you're out.

I took a trip to Norway last winter and made damn sure

I had plenty of Copenhagen with me. I've tried Norsk *snus* and don't like it. I happened to mention to a friend in Oslo what I'd done and added that to take snoose to Norway sort of struck me, as the saying goes, like "carrying coals to Newcastle." He laughed, and then he made a remark I hadn't heard before. "Yah, only here we would say, 'like taking cod to Lofoten'."

It's kind of sad snuff wasn't invented in Norway, but be that as it may, I sure think it has influenced a certain segment of Norwegian music. Particularly drinking songs. Many a time, booming out from the old Valhalla saloon or the present Cort Bar in Big Timber, I have heard that lusty old ballad, "The Battle of Copenhagen." Traditionally, naturally, on the *Settende Mai*—Norwegian Independence Day, the seventeenth of May, which commemorates Norway's freedom from Swedish rule. One verse, especially, has always enthralled me:

"A t'ousand Svedes ran t'rough de veeds
Pursued by vun Norve'ian.
Dey knew dey'd loose, for dey had no snoose!
At de battle of Copenhagen."

Music aside, snoose has had its part in international events. And rightly so!

13
THOSE PY WHITES

MR. MARCUS SNYDER was a good man. His dad was Captain D. H. Snyder, old-time Texian, Indian fighter, cattleman and trail boss. Mr. Marcus was cut out of the same rawhide, and he'd added oil to cow critters. When I knew him, he'd moved up to Montana, had a house in Billings and was running stock on the old Senator Kendrick outfit southeast of Hardin. I got acquainted with him through Big Bill Eaton. Matter of fact, it was Bill that located the first Quarter stud I ever owned, a Blackburn colt that I bought from Mr. Marcus. After that, he and I got to be pretty good friends, and I'd make a point of saying howdy whenever I was in Billings.

It wasn't hard to hunt him up, for he was in town most of the time while one of his sons, John, tended to the ranch. He had an office in town, but invariably he'd be ensconced in the lobby of the Northern. Hell, whenever he's in town, every Montana or Wyoming rancher will end up at the Northern. It's been that way ever since I can remember— and I can remember back to the Billings Fair in '19—rode a race or two for Dad during it.

Now Mr. Marcus was, like me, a conversational cusser, only he, being lots older, having covered more country and having had more practice, had a magnificent command of lurid language that put me plumb in the shade. I never inquired, for I was damn respectful around the old gentleman, but at some time in his long career I'd bet he had

162

whacked bulls or packed mules—maybe both.

Also, he was a little deaf, so he had a hearing aid. Not one of those trumpet deals that looks sort of like an old phonograph horn, this was modern with batteries and wires. And, just like every rancher I know who has tried a hearing aid, the damn thing didn't work to suit him. He sure used to get annoyed—complained that it squealed something fierce when he had it turned up and that in a room full of people talking it made such a racket that he couldn't hear anything but it. The lobby sure as hell wasn't a real shiny place to use it, naturally, so as we talked, he'd get progressively madder, with the result that he'd keep shutting the rigging down more and more. Must have been hard on the old gentleman, but I couldn't help getting tickled with what it led to. I suppose it was so he'd be sure of what he was saying, but the lower he turned his outfit, the louder he'd talk. So, pretty quick, you could hear him all over the room, gaudy words and all. Consternation gripped the lobby—the management, that is—everybody else grinned in appreciation—when Mr. Marcus got strung out with one of his old-timey stories. He was a damn fine old man, and I liked him a lot.

I was talking to him one day when he suddenly said, "John's cryin' about my brood mares an' th' grass, so I'm goin' to have to sell 'em. Won't split 'em up, though. Don't want ever' sonofabitch on th' Big Horn ownin' a Fishtail mare. 'Druther have 'em all go to one outfit. They're bred right, an' I want a horse man to have 'em."

My ears pricked up, and I went to asking questions. He'd have around twenty-five, maybe thirty. He named the bloodlines, and my mouth watered, so we got down to agreeing on a price and the date he'd have them gathered. When the deal was made, I guessed it wouldn't break, though it would bend the hell out of me, but I chirked up when Mr. Marcus said, "Son, one thing I've sure found out in my life. Th' goddammed cattle might be where you couldn't give 'em away, but a good horse, now, you can always sell a good horse. I've gotten by on 'em a time or two when they wasn't anything else."

That's crowding thirty-five years ago, but he was right. Damn if he wasn't.

Along in April Mr. Marcus sent me word that the horses would be in his ranch corrals in three days, so I called a trucker in Hardin, and Barbara and I arrived at the ranch early in the morning three days later. When I looked over the pen fence, it spooked me. There was no way I could buy all the horses I saw! To make matters worse, Barbara climbed up beside me, gasped and said, "Why, we can't pay for all those, Spike."

"Goddammit, I know that, honey."

"What are you going to do?"

There was only one thing to do. "I'll tell Mr. Marcus so."

I collared him at the chute where he was talking to the brand inspector. "Mr. Snyder, there must be damn near twice as many head here as you said you'd have."

"Well, I told th' boys I wanted 'em to comb th' heads of th' draws, an' it looks like they did. Why, I haven't seen that ol' gotch ear Billy mare since I turned her loose, fresh from Texas." As happy as if he'd run into some old friends. "An' those two bayo coyotes an' th' grulla. Got some age on 'em, but they sure have raised some fine colts."

He was fixing to tell me about some of the others when I finally got a word in, "I haven't enough money to pay for all those ponies."

He sized me up quizzically. Then, "How much you got?"

A little quick figuring. The amount agreed on, plus the trucking, would about clean me. But, hell, maybe I'd hit a lick rodeoing this summer, so I took a chance and named what I knew it was safe to write a check for. Still a damn long way from what he'd asked per head, what with all the extra horses, though.

"Hell, Spike, that'll do," with not even a hesitation. "Let's start gettin' 'em loaded."

So that was that, though Barbara looked pretty dubious when I told her what the tariff had been.

When the last load was being sacked up, Mr. Marcus said, "Whoa up a minute. John, unsaddle ol' 'Cannonball' an' put him in that truck," and when his son objected, "Git to it. He an' I have seen a scope of country, an' I want Spike to have him. He'll take good care of him."

And so, that evening I was fifty-odd mares and one gelding richer, and a damn sight poorer when it came to a bank account.

They were a good string of horses. They looked rough, so, since I knew I couldn't run them all, I started getting them looking a little better. Bob Langston was working for me that spring, and the two of us sure earned our bacon—and some horse tracks, to boot. The older mares were fine ladies, once they remembered what a man was. But some of the young ones were wolves, and I doubt they'd ever seen a human before except at a distance. And a long distance, at that. In time we had them pretty well ironed out. At least, they wouldn't bust the fence or take us when we got in the corral.

All but four, that is, for in the deal I'd fallen heir to four white fillies—albinos really—that were about as juicy stock as I've ever run into. Seems they were Mrs. Snyder's, at least they were carrying her PY on their left thighs. How they were bred I have no idea; though as an educated guess, I'd say a rattlesnake, a wolf and probably a scorpion or two were somewhere amongst their folks, for sure. Godamighty! They'd put a man out of the pen and dare him to come back in. Ahorseback even. It was worth your life to get into the corral unless you sort of diluted them with some that were gentler. So Bob and I, after we'd each gathered an assortment of scars, got pretty handy at getting between corral poles in a hurry. Outside they were just wild, but in a pen they were green-eyed boogers. So I said to hell with it, and we quit them.

We finally got all the rest handled. More or less. Among them was a grulla filly with some age on her who was just a shade below the whites. After she'd been snubbed a time or two, she'd stand, but neither end was what a man'd call safe. Even passably so. We named her "Blue Hell," and I mean she earned it. The whites we called, collectively, "those PY bitches." Wasn't the only thing we called them, just the nicest. And, by God, we never turned our backs on any of the four. Ever. It got so that if somebody yelled, "Look out!" around the corrals, they emptied right now. Of humans, that is.

Barbara used to come over most afternoons to see how we were doing. One day the boogers weren't in the main pen, so we tolled her in, and Bob started pointing out the different stock and telling her all about them. While she was engrossed, I sneaked up behind her and poked her between the shoulders with a stock whip. Bob let out a yell and made a run for the fence. So did she, and I've never seen her go over a corral as fast, though some rank old mamas during calving have hurried her a number of times.

It was a hell of a joke. Until she swung around and asked politely, "How'd you two jokers like to do your own cooking?"

We looked sheepish and mumbled around apologizing, but it didn't work. She had the last word, "You damn sure will be, one more smart deal like this." So we were real careful when she was paying us visits from then on out.

The grulla wasn't coming along worth a damn—had gotten to where she just sulled up, squealed and waited for either of us to get within reach. So I threw her in with the PYs, and Bob and I moved them all to a section across the Sweet Grass for the summer. Handy to the rodeo arena if we had to corral them. We mentioned what we'd done to Barbara that evening. In open-mouthed astonishment she said, "Moved them? By themselves? How in the world did you get them across that noisy old bridge?"

"Carefully," was Bob's answer, before I could say anything. *"Real* carefully," and he sure was telling the truth.

The old well-bred mares I threw into my brood mare string, we broke a few of the young stock I especially liked and by the end of the summer most of the rest were gone, sold into the Dakotas, Canada, Washington and Montana. I never mentioned the renegades across the creek to any of the buyers, for I didn't want to spook anybody. I doubt we could have gotten a good look at them anyhow, for we never saw them ourselves except at a considerable distance and along towards dark. Looked like they brushed up in the draws during the day and grazed at night, like deer.

Well, in October I got a call from a couple of men up at Bynum. They'd seen some of the stock from my place and wanted to know if I had any more for sale. I told them I was about out of horses to sell, but, figuring maybe, just maybe, I casually mentioned that I did have four whites and a grulla, all fillies, that I might be persuaded to let go. I don't know whether it was the sex or the color, but be damned if they didn't get interested and want to know all about the five. Though it might be questioned in some quarters, I *do* have a modicum of decency, so I explained that these were broncs, had been worked with a little but were pretty wild. Which was the truth, but just barely. I added, too, that the grulla—whose name was not mentioned—was sired by Texas Bluebonnet, an own son of Joe Hancock. Mr. Marcus had used him for several years, he was one hell of a breeding stud and had just lately been sold to the Hayes boys down at Thermop'. So the upshot was that the two would come down and have a look at my treasures.

Which meant we had to get them back across the creek, for there wasn't a loading chute at the rodeo arena in the unlikely event that these two men would want them. Bob was gone by then, so after the two older children were off to school, Barbara and I saddled up. Shelly was too little for school, but we couldn't leave her alone in the house, so we put her on old Mickey Mouse and headed for the section west of the Sweet Grass.

I'll never forget that day, nor has Barbara. Shelly, too, for I mentioned it to her not long ago, and she grinned from ear to ear, "That was fun. I thought it was a spectacular way to move horses." I can't agree, even looking back thirty-five years. But it was sure enough spectacular!

We located the five, I eased them out of the draws over south and they disappeared over the hill like wild animals. Barbara was there, mounted on Sleepy, a running thing, and as I topped the hill, I saw her turn them down the north fence. I spurred for the east fence, and time I sighted it, the five were in overdrive along it headed south. I bent the lead, the grulla and the youngest filly, but the other three split up and went by me like I was standing still. I outran them, but when I tried to turn them, they flat ran over me. Pretty quick Barbara loped up and asked, "Where'd they go?"

"Back where we started, I think. Either that or to the Yellowstone."

"Well, not all of 'em. Blue Hell and one filly boiled by me at the north fence, and I turned them up into the northwest corner. They're there now. Probably catching their wind."

While we were talking, Shelly arrived. She'd had a nice run, and her eyes were sparkling, but she was on the fight because we hadn't waited for her. So we cooled her down and started another go-round.

Once again I brought them out of the draws, they joined their compatriots in the far corner of the section, and we started easing them down the fence. The gate out of the pasture was in the northeast corner, and I damn sure had it open and ready.

I've moved antelope from one pasture to another a time or two with Dad or Tor. It can be done if you are really ahorseback, have your gate open, keep far enough away so they don't cut back but close enough to keep them moving. But these damn horses were wilder than antelope. About two-thirds of the way down the fence they broke between us in spite of everything we could do. This time we lost the whole works, so when we regrouped no one had much to say. Out loud, that is.

The third try wasn't a damn bit better, and as the five disappeared over the south skyline, my wife asked bitterly, "Got any fresh ideas about this horse roundup? Real fresh ones?"

"Yup, I'm about to ride home and get my rifle. By God, but I'd love to down that oldest white slut just about the time she throws it in high gear. She's the troublemaker."

"No, you can't do that," she argued, "if we can't get them in where these men can see them, let alone load them, we sure can't sell them. And what'll we do with them if we can't?"

"Then I can have the pleasure of shooting them. All of them. You know, damn if I don't think the bastards are enjoying this."

Barbara smiled tightly, "Well, I'm glad somebody is, besides Shelly," for the latter had just arrived with a grin like a skunk eating garlic. "Gee," she greeted us, "Mick can run, but she plays out. So I've been resting her on the way over to you."

Well, we had a fourth go at it. A little better this time, for we got them almost to the top of the hill above the gate. Then they broke. Came at me in a bunch, and I wisely got the hell out of their way before they ran plumb over the top of me. As they came past, laid straight, I was astounded to see Barbara right on their tails, her big running mare wide open. And she was screaming every jump. It spooked me, so I fell in after her. At the ridge top she pulled up and jogged back. When I met her, one look at her eyes warned me to be very, very discreet. So, though I was busting with curiosity, I just asked, "Are you all right, honey?"

I could see her relax a little. "I'm fine, just fed to the eyeballs with those five, damn them."

It was safe now. "What was all the noise about?"

"Why," and her eyes warmed, "if those sonsofbitches want to run so badly, I thought I'd give them a reason to. Wish I'd had a bull whip when I rode up on them."

Wasn't much I could say to that, and I agreed with her anyhow. So we blew our saddle horses until Shell showed up. At a trot this time, for old Mick was getting leg weary.

We decided on one more try. If it didn't work, we'd go back and bring over a string of gentle horses. Maybe we could get the ridge runners to stay with them. I had an idea that just might work first though: we'd use Shelly and Mick as decoys. We would put her where the north fence topped the hill above the gate. Then, when we had the horses lined down the fence again and they could see her, she would ride down and through the gate.

Barbara was dubious as all hell about the plan. After all, Shell was only five. But she was a hand! So she promised to go through the gate well ahead of the boogers and to get out of their way when—and if—they came through. "Yes, but what happens when they are through?" her mother asked skeptically.

"The high rim along the creek'll stop them from going east. Anyhow, they'll turn south again on the inside of the fence. It runs up against the arena, the gate's open, and if Shell will jog down to the corner at the far end of the arena, we've got it made. I'd bet on it."

Damn if it didn't work like a charm. The first part, anyhow. Maybe the renegades were run down a little or something. We gathered them again, and Barbara brought them down the fence, laying back as far as she dared so she wouldn't spook them, while I rode well off to one side of the point. They spotted Mick—I doubt they even realized Shell was on her—followed her through the gate and busted south along the fence. We both hit the gate at a hard run. I closed it, and my wife said with a beatific smile, "Thank God." I agreed wholeheartedly.

I'd barely gotten ahorseback again when here came our pets along the fence towards us looking for the gate they'd just come through. We headed them and started back for the arena, real easy, with me holding well to the east. They got to the arena, and the grulla ventured in, but the whites spooked, and out she came, too. This time they were ranker to bend, and I had my doubts about whether we could get it done again. So, as they neared the arena at a lope, I yelled, "Crowd 'em," and closed in. We hit the gate at a hard run, and in they went. All but the oldest white. She jumped the fence into the section we'd just gotten them out of, slick as a deer could have done it. Never touched it. Oh, how I ached for my rifle as she headed for the hills! Not running, just loping proudly, head and tail in the air and snorting like a buck antelope.

I hurriedly shut the gate and tied it tightly. Then, when I could trust myself, I turned to Barbara. "Did you see that?" I asked incredulously, just as though she hadn't been right beside me.

Here came Shelly before she could answer. "Hey, Dad, what happens now? Do we get to chase her again? I hope."

"No, Shell, we'll go over and get some of the horses at the barn and see if we can get these four over to the corrals at home with them," and in an aside to Barbara, "I'll pick up a rifle when I go by the house. I've had it with that white bitch."

While we were gathering the gentle stock, I cooled down a little, and the humor of the whole deal hit me so I left the rifle behind. We left everything open when we brought the horses over—the corral gate, the gate from the lane into the corral lot and the lane gate into the arena pasture. We threw the stock we'd brought in with the wild ones until everybody got well acquainted. Then, with Shell in the lead with orders to take it easy, we started home. As we came out the lane gate to the bridge, Barbara said, "Look," and pulled up.

There, on top of the highest hill in the west section, a good half-mile away, stood our missing PY. Head and tail up, watching us, and we could hear her whistling! We watched her a minute, and Barbara said, "You know, Spike, she misses her friends. She sees us and the horses, so let's just leave this gate open and the one into the section. She just might follow, even over the bridge, if she gets lonesome enough."

Personally I figured there'd be as much chance of that happening as a snowball would have in hell, but I did as she suggested. Then we rode on and found Shelly had everything corralled and was holding the gate. The kids were back from school, so, while Barbara went to get supper, we cut the gentle horses out north, put the four boogers in the back pole corral and I double-chained the gate. Time I'd tended our saddle horses and chored and fed the captives a fork or two of hay, it was getting dark. Before I headed for the house, though, I propped the gate to the main corral open. You could never tell. Then, as I was walking to the house, I heard a horse whinny, 'way west of the creek. And every one of the four snakes in the corral answered.

The next morning when I went to chore, I got the shock of my life. There, in the big corral, fraternizing with her buddies through the poles of the back pen, was the PY ridge-runner. I didn't believe it—can hardly believe it right now—but I slipped up and closed the gate on her. Just in case she was real. Then I beat it for the house, got Barbara and told her. As we hurried to the pens, I heard her muttering something about my "seeing things." Then she peeked through the poles. Her mouth dropped, and she shook her head a time or two, turned to me and breathed a heartfelt, "Oh, Spike!"

"You were right, honey," I told her happily. "By God, we're in business."

We had three pastures on the creek bottom. Two opened into the corrals. The third was bigger and opened into the middle lot through a gate by the edge of the brush. The far fence wasn't too good, but the five were acquainted with things from last spring, the grass was fair and they'd be well away from anything that might stir them up. I wasn't about to throw them out in the north section. No way! So I opened up the two lots, cracked the corral gate into the middle lot and left the five to ease out of their own accord. By night they had worked into the far pasture and were plumb in the yon corner, keeping a wary eye on things. I shut the gate, and we left them there, hoping they'd cool down. Seemed to work, though any time there was activity at the corrals, the one white bitch would lead them all to the far corner and stand there whistling.

On Armistice Day the men from Bynum, Roland Peters and Clarence Kramer, arrived. Looked like they were live ones, too, for they were driving a good-sized truck, but we damn sure didn't let our hopes get high.

Barbara and I saddled up, toying with the blissful thought that this might be the last time we'd ever have to mess with our pets but still with the nagging apprehension that something bad would happen. I got the corrals all set, but before we rode off, I told the two men, "I'd be obliged if you'd stay out of sight while we're penning these ponies. They're pretty ouchy, and I'd sure hate to have them go through a fence. We'll be right back."

Of course, the five had spotted us the minute we showed up at the corrals and had high-tailed it to the corner. We worked up on them casually as all hell—which was a far cry from how we felt—and they dropped down toward the gate, old PY in the lead. As they headed for the open gate, she reminded me of a coyote slipping up on a trap he knows is there but can't quite spot. My saddle horse began to get uneasy—her boogery snorting had him wondering if whatever it was might get him, too.

They were just about at the gate with us easing along behind them when Barbara, in a small voice, said, "Heavens!" and her face turned white. I followed her gaze, and my reaction was "Good Godamighty!" but I said it quietly. Real quietly, for just off to one side of the gate were our two horse buyers, hunkered down in the chokecherry bushes. They were "out of sight" all right, as long as neither of them twitched a muscle. If either did, there'd be a big hole in the north fence and another section or so to gather. We didn't dare stop, though hell was on the absolute verge of popping, but I be damned if PY didn't lead those wolves through—strung as tight as a banjo string, stepping like she was on eggs and rollers in her nose that they must have heard up in Melville. As soon as they were by, the two men stood up. PY reacted as though she'd been shot in the rump with a red-hot shinglenail. She'd spotted the corral gate, though, and the five poured in, on into the back pen, we slammed the gate and allowed ourselves a moment of thanksgiving.

Well, we sold all five; Kramer took the whites and Peters the grulla. The price sure wasn't top-heavy, and I might have gotten a little more, but hell, I'd have been glad to *give* them away. Anyhow, one of the most satisfactory moments of my life was when that truckload of horses, by the grace of God all uncrippled, pulled out of the ranch and down the road out of sight!

When I got a letter, postmarked at Bynum November 17, I was a little leery of opening it, but finally my curiosity got the best of me. It was from Clarence Kramer. He started it, "Dear Friend," which was damn encouraging. If he'd said, "You sonofabitch," I wouldn't have been surprised. I read the letter, reread it and reread it again. Then I showed it to my wife. Her reaction to it was the same as mine. We were both absolutely thunderstruck. How in God's name had he done it? I've kept that letter all these years, and here it is, word for word.

Dear Friend;

Well, we got home in fine shape with the horses, that is, to the stockyards in Bynum. Then the next day we took one of Peters' mares up and put her in the yards with the new horses. We then turned them out and started to chase them home. They had to cross a creek, and they were spooked by the time they reached it. Just as the grulla was just about to step in the creek a chink flew up right in front of her. That started the fireworks, they ran thru a fence, the albinos got away for the time being. Grulla and Peters' mare ran together so we after about 2½ hours chase, we got his horses in his corral. The next morning we started after the albinos, and you did not lie one little bit when you said they could run. We relayed them with three saddle horses for six hours that day, and I finally got my rope on the two-year-old filly, then she killed herself before I got her home. The next day we got the other three corralled at a neighbor's place about two o'clock. We roped two of them from the ground and got the halters on them. As luck would have it, I roped the PY four-year-old from the saddle horse. She never tightened the rope. She turned around, started to come to my horse. A neighbor boy stepped out from behind my horse and did she take him in a hurry, her mouth open and both front feet flying.

She chased me out of the corral twice yesterday, and my boy once. But last night I rode her around bareback, today I rode her with the saddle, but she is not to be trusted yet. The other two are awful loving once you get them up to you.

Yours truly,
Clarence Kramer
P.S. Boy are those ponies classy under the saddle.

I answered him. I told him I'd lost a good colt myself—though I didn't say it was my own damn fault, which it was. I also remarked that I didn't imagine either he or

Peters felt real neighborly towards me when they were trying to corral their purchases. I got another letter from Clarence, mailed December 2. I've saved it, too. He was quite a guy.

Dear Spike;

I am sorry as hell to hear you lost your good colt. It sure takes the wind out of our sails to lose horses like these. It is a good thing we can raise more.

No, Peters nor I cussed you atall. We just decided that the next time you said a horse was wild, we would believe you and keep them in a corral until they were gentle. I have not seen Peters for a week or so, but the last time I talked to him he said he might not try to break the grulla until she has a colt or until he can see if she is bred.

These pictures will tell you more than all the writing I can do. I have not rode the ponies enough to see if they will singlefoot naturally yet, but they sure walk as though they were on eggs. Velvet is getting so she will come to me now, but lord how she can snort while she is doing it. I don't know whether Milky was broke to ride, but she sure as hell was broke to lead on foot or horseback. She never tightens the rope, no matter how fast or slow you go. I had the saddle on Milky Sunday and rode her around the corral for about twenty minutes without any trouble at all except she had a notion to kick me when I got off. I can handle her pretty well now if I don't have anything in my hands except the lead rope. Whenever she acts as tho she is going to cause trouble, I tangle her all up in her rope and down we go. When she gets up, she behaves like a little girl for a day or two. I stuck my handful of oats in front of her yesterday, and she opened her mouth as tho to bite my hand. Then turned her head to one side as tho ashamed of herself and came back for her oats real gentle like, so that gave me some encouragement which I was very much in need of about now. A neighbor of mine was here Sunday when I was riding Milky and was he taken to her. He remarked about her walk, too.

If anybody down there has any money that says those whites can't be broke, go the limit, for I am positive it can be done. Velvet is real loving, as the picture will show, after she gets close enough to lay your hand on her forehead. The Baby Sis is crippled on three legs at present, rope burns and bruises, but I think she will be alright in a month or so.

We had a couple of weeks of damn ruff weather when the horses were neglected a lot so I lost a lot of ground with them.

So long for now and good luck.

Clarence Kramer

I've forgotten, though he and I did quite a bit of horse business in subsequent years, whether Peters ever got Blue Hell broken to ride. I know she made him a good brood mare, but I think she ended up as a saddle bronc—finally broke a leg, I believe, at a show in Choteau. Clarence gave the Baby Sis to his boy, and she never was really broken. But Clarence broke Milky and Velvet and used them for a long time, though I've heard rumors that he was the only damn man in the Teton country that could ride them. Hell, I'd have bet there was *no* man could have broken them. Anyhow, in right at seventy years with horses I would have to think real hard to come up with any ranker stock than those five.

By the same token, I've known a lot of horse hands; some thought they were and had their shingles out as such, and some damn sure *were*. Clarence Kramer belongs in the last bunch and right up towards the lead. By God, he broke those PY whites!

"Christmas at the Line Camp"
Courtesy of The Charles M. Russell Museum,
Great Falls, Montana

14
CHRISTMAS AT THE SWEET GRASS CAMP

Nearly twenty years ago we sent Barby down to Omaha for her freshman year in college. She was a pure quill ranch girl and, naturally, terribly homesick. To make matters worse, we weren't very flush. When Christmas

vacation rolled around, we just flat didn't have the money for her to make a
trip home. In hopes that it might brighten things, I sent her the Charley
Russell card, "Christmas at Line Camp," and I wrote her the following letter.
She has kept it ever since, so I guess it worked!

December 25, 1963

Hi there, Honey:

Me an' Rimfire's holdin' down the Sweet Grass camp, the weather's been salty, an' while we got plenty of beans an' venison, we're down to the last can of Carnation an' got about a half a sack of Durham betwixt us. So this mornin' Rimfire says, "I think I'll just saddle that snorty bay an' make a trip into Melville an' see if I can get enough canned cow to do 'til the boss comes 'round, an' some smokin'. He needs a hard ride anyhow, damn him."

Well, I opens the water for them thirsty cows, hooks up Bullet an' Babe an' hauls a couple of jags of hay, fills th' woodbox, an' along in the afternoon slips out an' knocks over a fat dry doe, figurin' fresh liver'd go good when Rimfire gets in. She's anyhow a thirty mile round trip to Melville, an' cold, I'll tell a man!

'Long about the time the sun's goin' down through the frost haze up the canyon, he comes in, an' he's outa sorts. Seems them boogers layin' around the YMCA in Melville about eat the store outa grub, an' they's no milk nor tobacco left. "I did manage to talk Hanson outa a little somethin' though," Rimfire says, an' pulls a bottle outa his coat. He's all right too, the seal ain't broke yet!

He tends his pony while I gets some heart an' liver on the stove, an' after we eats, we settles down with the bottle, stretchin' her out with coffee, which we got lots of. When it gets down towards the bottom, Rimfire says, "Maybe we better save enough for a short one in the mornin'."

"Fine," says I.

"By the way, Spike," says he, "you got any idea what day it is?"

"Nope," says I.

"Well, if them fellers in Melville's right, tonight's the night the kids hangs up their socks."

"Sure enuff," says I, "I'd be 'shamed to hang mine up where anybody'd see 'em, besides I ain't no kid. Reckon I'll turn in."

Well, I'd drunk a lot of coffee along with that whiskey, an' figures maybe I better have a plumb early mornin's mornin' before I hit them sougans, so I throws my coat over my shoulders, steps out the door an' there he is!

Kind of sets me back for a minute, but wonderin' just what calibre that booze is anyhow I ups an' says, "I know it ain't polite, but who're you?"

"Claws," says he.

"I knowed a Claws when I was ridin' for the Three C Bars on the Upper Milk, Cuts the Bears Claws, but from the looks o' your riggin' you ain't no kin."

"Nope," he comes back, "the first name's Santy."

"Why sure," says I, "you used to come around back home when I wasn't much more'n a weaner! Look, why don't I help you hobble them bucks an' you come on in an' warm up an' eat?"

"Thanks, son," says he, "it's been right brisk, an' I'll do just that. They stand good though."

When we gets inside, the ol' feller knocks the frost off his whiskers, sheds his outfit an' backs up to the stove like a sick kitten to a hot rug. I puts the skillet on whilst Rimfire pours what's left of our whiskey into a cup along with some water from the kettle an' hands her over.

"Obliged, son," says Santy, "but ain't you two goin' to join me?"

"Nope," says Rimfire, "we had our'n. Drink up."

He does, an' then pulls up to the table, an' it sure looks like it's been a long time betwixt meals! When he finally shoves back, I pulls out our smokin' an' passes over the sack. He grins, gets out a runty pipe, fills her an' lights up.

Well, the ol' feller sets there apuffin' an' anoddin' whilst us two breathes deep so's we'll get a little somethin' anyhow, an' watches the light from around the stove lids dancin' on the ceilin'. After a spell, Santy straightens up an' knocks out his pipe. He gets into his coat an' mitts, pulls down his cap an' says, "I'm sure obliged to you boys, but I got a long ways to go. Thanks, an' a Merry Christmas to you both." An' then he just flat ain't there no more!

I looks out the door, an', by God, there ain't no cutter, bucks, nor even tracks! Just a coyote singin' to the moon up on the Basin Creek ridge. Behind me I hears Rimfire say, "I be damned."

"Me, too," says I, "reckon we're drunk?"

"Maybe so," says he, "but don't look a gift hoss in the mouth. See yonder." An', there by the chair where Santy'd been, sets a couple of caddies of Bull Durham, a case of Carnation, an' two quarts of booze.

We looks at it, each other, grins an' says together, "Merry Christmas!" An' I swear them same words echoes back out o' the night, 'way an' away, along with the chimin' of bells!

Love,
Dad

View of Melville Rodeo,
August 1935

Dude fun on a Sunday afternoon

Some old-time cowboys

Sunday morning meeting!

Youngsters!

Getting the kinks out

Some Lazy K Bar horses

Early day Melville

The Fischers

Two Dot Wilson with the gallery

Brannin family

Brannin cabin/home

Some of September Morn's relatives

The stage stopped at Bert Blethen's blacksmith shop

Columbus, Montana rodeo

"Dad's Home"

At the Fischers' golden wedding celebration

Butte Ranch—main house

Butte Ranch

192

Big Timber in 1886

F. Jay Haynes
Permission to reprint courtesy of Mrs. Jack Haynes

194

Frosty Johnson's camp

Stacking hay

195

Start of Indian relay race at Billings   *Stations on a relay race*

Columbus, Montana rodeo

Melville Rodeo, 1936

Getting ready for the Big Timber rodeo, 1925

Spike's Grandchildren

*Tony Carroccia,
1981—at the ranch*

*Shannon Kirby, 1977—first horse show and first ribbon on
Sam Hill, Big Timber, Montana*

198

Rocco Carroccia, 1981—with his first elk, a 6 point bull.

Kelly Kirby, 1980—with Paleface, a fine old kids' horse

J Carroccia, 1975—in the Big Timber parade

Page Carroccia, 1980—with her colt, Cinder

═══ 15 ═══
PUT A KID AHORSEBACK

WHEN I WAS A YOUNGSTER, there sure were a lot of things that were supposed to be "good for me." Dad in particular seemed to figure, "Put a kid on a horse—and use him." And then, he'd qualify it with, "He'll learn responsibility. Be good for him." So, especially if it wasn't handy to spare one of the crew, I got pressed into service for quite a variety of jobs around the ranch. Most were ahorseback which suited me, though I didn't always get to pick what I rode. Come to think of it, it probably accounted for my cavalier treatment of my own children years later. We all survived, though.

There was one job I drew fairly often which was not my favorite. Dad had a section on the west end of Porcupine Butte with the only water on it a sorry spring that invariably iced up in real cold weather. We used the section mostly as winter range because there was always wind whooping down through the American Fork pass and it kept the west side of the butte open. But the stock water was something of a problem. When the weather got real cold—twenty below and more—somebody had to ride over and "chop water" about every other day. Being in school, I was only handy on weekends, but it was funny how often that damn spring had to be opened on a Saturday or Sunday. Hopefully, there might be a chinook during the week to save my bacon, but as Gramp claimed, "The only way to be sure of getting a good chinook is to start putting

199

up ice." We only did that once a year so it didn't help me much.

It was six or seven miles from home to the spring, for I was told to stay in the county lane to the Mydland crossing on the Sweet Grass so they'd know where to look for me if I got in a jackpot. From there to the spring wasn't bad, but the lane from Basin Creek to the crossing, a straight run of close to two miles across an open, rocky flat, was a booger, and whenever I hear the work "bleak," to this day I can see that long, lonesome expanse—and its weather!

More often than not, there'd be a ground blizzard—right in my face on the trip out—maybe up to my horse's belly or shoulders, so that the whole flat was moving. A wide, white river, almost a sea; the hills in the distance stuck through like islands. Watching everything stream down and past made a man dizzy. If the wind were hard enough to really move the snow, it would be God only knew how high, and I was lucky to be able to make out the fence posts beside the lane as my mount trudged by. I'd have to keep my face tucked into the collar of my sheepskin coat and the brim of my muskratskin cap pulled over my eyes to save them from the lash of the snow.

Or perhaps there'd be a vicious drift of air out of the northeast—which I'd have to quarter into on my way home—with the sky and ground completely indistinguishable. Scurrying little trickles and snakes of snow would appear in the white ahead, whisper past and disappear into the white behind us. Only the Crazies and Porcupine showed their tops above the frost haze—unreal, bitter and icy blue.

Or it might be clear and so quiet I could hear sleigh bells a couple of miles away as a neighbor fed his stock and always the creak and squeal of my horse's hooves on the dry snow to keep me company. A glittering day with the snow to the horizon sprinkled with points of light and the hollows in the shade a deep, deep bottomless blue. If it were

cold enough, and it usually was, on a day like that there'd
be sun dogs: sullen orange rimmed with gold, flanking the
sun, and the whole sky would be filled with the diamond-
like sparkles of innumerable frost crystals. Lovely, but
damn hard on the eyes. Nobody'd heard of sunglasses back
then, so when it got too bad, I'd tie a thickness of light silk
bandana over my eyes to cut the glare and still let me see.

Naturally, I had to have an axe to open the spring, and
the simple expedient of leaving it where it was used until
we didn't need it any more didn't fit into Dad's scheme of
things. "Might drift under" or "Somebody'd have to make
a trip over for it in the spring." So I carried it each trip.
Both ways.

On frozen ground with an unshod horse—Dad didn't like
to sharpshoe a saddle horse, claimed it made them too
brave and careless—it isn't real smart to tie an uncased axe
to the saddle because if a horse should slip and go down or
turn over, either he or his rider might get badly cut. Since I
never even heard of an axe case until I got around Forest
Service trail crews a number of years later, I had to carry
the damn thing by hand. And if there's anything more
unhandy than an axe—or colder, even with mitts on—to
carry ahorseback, I've never run into it. I tried everything:
first one hand until it got cold; then the other; across the
saddle, but it kept sliding toward the heavy end unless I
held it tight; on my thighs; on my shoulders; and even
across the cantle and sitting on it. I quit the latter after one
try though, for an axe handle can be colder than hell where
Angora chaps don't cover and a man can't even warm up
his cantleboard with a chunk of wood between it and his
seat. So it was just a case of continual shifting, and when
the hand I was holding it with froze up, wrapping the reins
on the horn—my horse was gentle and knew, only too damn
well, where we were headed—swapping hands and beating
the cold one on my leg until it got some feeling back. Then
it was back again and work on the other hand. I don't

suppose the thing weighed over five or six pounds, but by the time I got home, I'd have sworn it was at least forty. Gramp had read me all about Sinbad, and I sure sympathized with him and his Old Man of the Sea. I believe the word is incubus!

Opening the spring itself wasn't much of a job and didn't usually take long. Trouble was, it was in a pretty steep coulee so I had to try to chop loose the ice around the water on the slope so some fool cow wouldn't slide in on her head and either drown or get down. In the latter case, it was a cinch she'd be dead in a couple of hours, the rest of the stock couldn't get to the water and she'd have to be skidded out of the way the next trip over. There are easier things than cutting a frozen carcass loose from frozen ground and dragging it away, particularly with a barefooted horse. So a man was damn careful to roughen what ice around the water he couldn't chop, then cut some chunks of dirt loose wherever he could on the hill, pound them as fine as possible with the back of the axe and scatter the results over the packed snow around the spring so the stock that came to drink could get some footing. I guess I must have done all right, for we didn't lose any cows, but to be truthful, I wasn't nearly as worried about a cow dying as I was about what would happen to me if one did.

Then onto my horse again for the trip home in company with that damn axe. Once in a while we wouldn't get to the ranch until after dark, and those hard, steely winter stars are a far cry from their friendly summer counterparts. All in all, I guess, miserable as those trips invariably were, Dad was right: I had my doubts at the time as to just how "good for me" they were, but I damn sure learned responsibility. Possibly, too, a contributing factor to my admiration for horses is the fact that a man never has to "chop water" for them. All they need is snow, thank God!

I won't forget another little soiree I had at Dad's behest. I was older then, just at the age to figure I was a cowboy from away back, and though the facts have never been made public, it was nevertheless a deal that made me squirm badly.

Dad had sold an old gentle mare for an irrigating horse to a rancher just across the Yellowstone from Big Timber. She was up at the canyon ranch, so he decided he could spare me at the corrals for a day to deliver her. Since he didn't want the trouble of picking me up, I was told to lead her down and ride home. Suited me fine, especially the horse he told me to use: a big, classy colt that had been broken the year before and a traveling thing. I'd ridden him a number of times and liked him. So I saddled up, haltered the mare and away we went. I must admit I showed off a little for the dudes when I got aboard. Cheeking him and pulling him around into a tight circle when my pants hit the saddle, all of which I've learned by now is not very smart—if you treat a horse like a bronc, he's liable to get to thinking he is and is damn liable to act like one.

We made good time, but the mare had trouble keeping up with my pony, so most of the time she was back at the end of the lead rope while he was swinging along with his head up and his tail in the air. Just the horse for a booger of my caliber. Sort of a shame I hadn't met anybody.

We dropped out of the mouth of Plaggemeyer Gulch and turned south down the lane towards Big Timber Creek. It was hot, not a very exciting stretch of country at best, so I was sitting all gapped open and paying damn little attention when the mare, tired of getting her neck stretched, trotted up on my near side. The minute my horse felt the lead rope around his hind end, that proud tail clamped down on it like a vice. He made a jump, the mare set back in surprise, the rope burned under his tail and it was Katy bar the door.

I was hovering at the ends of my reins before I knew what was going on, and when I did, it was to realize I was flat on my back in the middle of that hard road, wind knocked out and the dust boiling around me. I got shakily to my feet and when I could see and breathe again, made my tally. It wasn't good, not even discouraging; just plumb, by God, humiliating!

There I stood: booted; spurred, chaps, too; alone; afoot; my saddle horse headed for the Yellowstone at a run. In the other direction, the mare was going home at a long trot, her head off to the side so she wouldn't step on the lead rope. Talk about a lonesome feeling. But, damn, I was glad I *was* alone—nobody'd seen the proceedings.

If there was the chance of a snowball in hell of catching either of the two, it'd be the mare, so I started trudging north. I'd made maybe a half-mile, the mare long gone out of sight, when over the rise ahead of me came a wagon. As it got closer, I saw it was Oscar Vik, a neighbor from up on Big Timber Creek, and, joy to behold, tied at the tail gate was my missing mare. Oscar whoaed up, grinned and asked, "Troubles?"

"Sure do." Then, realizing that I hadn't dusted myself off, I stifled the urge to give him some classy story about how come I was afoot all by myself and added "Thanks for catching her. I've got another one up ahead, too."

"Yump on, and ve'll go see." No questions; damn, but I like people who mind their own business!

Surprisingly, when we crested the hill a mile or so down the road, there was my horse. He was young, in strange country, knew home was back of him, lonesome and damn near as glad to see another horse as I was to see him. We stopped, and I led the mare up in front of the team while Oscar slipped through the fence, eased around the colt and worked him quietly back toward the mare. When they smelled one another, they had quite a reunion, and in the middle of it I collared a bridle rein. I was back in business. Unbelievably.

We traveled together for a while with me keeping an eye on my lead rope as did my horse. Then I thanked Oscar again and jogged on ahead. I delivered the mare without any more excitement and cut across country for home. Over the ridge to Swamp Creek, across the Long Gulch to the Forks of Big Timber, up the North Fork, and I was unsaddling when the supper bell rang.

I never mentioned the day's incidents to anyone, especially Dad. I'm sure Oscar knew what had happened, but if he'd told on me, it's a cinch somebody'd have gotten on me about it. He was a nice guy. I can laugh now, but I can still remember how a cocky young buck, with quite an opinion of himself as a cowboy, felt. I'd have sold out cheap!

Another trip I made for Dad sticks in my mind for two things; the person I delivered a horse to was named Pearl Maddox, and he or she lived up on the East Boulder River.

The name "Pearl" hooked to a man—for I took it to be a man from what I gathered from Dad—fascinated me. I was used to average names for men, though Sigurd, Olaus, Halvor and others of Norsk flavor were pretty normal due to the promixity of "The Settlement." But, "Pearl"— maybe I'm not spelling it right, but that's the way it sounded—I sure wondered what he looked like.

Ranch kids in the Melville country were provincial as hell back in those days, and I was going up the Boulder—on the yon side of Big Timber and the Yellowstone! I'd heard of it but never had been there and wondered what sort the people were. Maybe like those on the Glasston Flats south and east of Melville—strictly farmers and guys who killed one another over irrigating water. They did, too, by God! Or maybe they were like the dry landers that lived over north of Fish Creek; Southdowns mostly, who were always squabbling and burning one another's stills. That is, if you could believe the stories told at the Settlement school. And we did, we did.

So I was eager to make the trip when I saddled up one morning and headed for the Yellowstone, though maybe a little apprehensive, to boot. I never did know why I was delivering the horse, but whatever the animal was, it must have been pretty gentle for I couldn't have been more than about nine years old.

I rode to Big Timber the first day. Not near as exciting as I was sure that Boulder would be, because I'd made the trip before during shipping time, but I enjoyed it. I knew the drivers of most of the rigs I met, or they knew me or Dad, and we'd always whoa up and talk a little. Naturally at some time during the conversations, I'd remark offhandedly that I was headed for the East Boulder with a horse for Pearl Maddox, but the results of this information were definitely disappointing. Nobody seemed to give a damn outside of a "Sure 'nuff," or a "That so?" and I decided probably they didn't completely understand just how damn important the Boulder was, anyhow.

Well, I hit Big Timber, put Pearl's horse up at the West Side Livery, went down to the Grand Hotel where Dad'd said to go and found they had a room for me. Then I ate at the Chinaman's, Tom Que's American Eagle, but never completely put my full weight down. After all, I'd heard stories—at the school, naturally—and from "Stampede," who worked for us and knew everything about everything, of how the Chinamen had all sorts of tunnels underneath the streets of Big Timber. Tunnels which culminated in opium dens where God only knew what terrible things went on. As a matter of fact, though, I had a fine supper for the vast sum of two bits, and Tom, who seemed a nice old guy, fit me out with an extra piece of pie. Funny thing, too, he wasn't a hell of a lot different colored than any of the hands I was used to who'd been out in the sun a lot. Remembering "Stampede," though, I decided Tom's eyes did look a little dangerous, at that.

Well, the night went fine, barring an electric sign somewhere outside my window which kept me a little worried about whether the hotel was afire or not. Still, there was a good stout rope with a knot every couple of feet fastened to a ring bolt under my window, so I figured I could escape if need be.

Breakfast at the American Eagle was good, though time I'd worked through a piece of steak and some eggs, I just flat couldn't handle all the hotcakes served me. When I got to the livery barn, my horse was saddled—at least someone savvied that I was on an important mission. Hell, I'd never had a horse saddled for me in my life up until then.

The trip up the Boulder rocked along in good shape. Pretty country—a wide valley with a creek swinging back and forth through it. Twice I had to cross bridges, and from them I could see that it was a bigger creek than any over home and rocky as all hell. But not as pretty water as the Sweet Grass. The ranches I passed looked about like those in the Melville country, and the people I met looked, to my disappointment, plumb normal, but the cliffs along the bluffs were different than our rimrocks. Years later when I studied geology, I realized they were made of conglomerate, "pudding stone." In the early afternoon I got to McLeod. There was a store there, but conditioned by my riding with Dad, I never even thought of any lunch. It would have been sort of fun, though, to get a bait of crackers and sardines like I'd seen riders do when they hit Melville; would have made me feel like a sure enough hand. Instead, I asked directions and headed on up the creek to the East Boulder road and turned up it.

The country opened up to a wide basin, and, though to this day—for it was gone the next time I was up the Boulder about twenty years later—I swear that there was a great big pipe running up to the top of the bluffs east of the creek. Couldn't have been for pumping water, for it didn't go all the way to the creek. Maybe it was for getting grain or some such down from the flats on top. In any event, so I wouldn't show my ignorance and give the Melville country a bad name, I didn't inquire about it from the infrequent travelers I met. They didn't seem to want to talk, nor were very friendly, and I didn't know any of them, anyhow. Especially the kids.

I finally got to the ranch I'd been pointed at. It was pretty well back against the mountain and looked about like any ranch up under the Crazies—all the roofs high-peaked so they'd shed snow well. A man came out, though I be damned if I ever found out whether it was Pearl or not. Matter of fact, I never did learn who Pearl Maddox was; never once was anybody spoken to or of as "Pearl." Maybe he owned the outfit; maybe he was the hired man; maybe, even, the woman in the house was named Pearl. To my intense disappointment I have never to this day been able to say honestly, "Hell, yes, I knew ol' Pearl Maddox."

I spent the night there. Slept in the same bed with a couple of kids about my age. They seemed nice enough, but we never really got acquainted—sort of hung back and sized each other up pretty carefully like strange pups. Besides, they were running barefoot, and while I didn't have boots, at least, by God, I was wearing shoes. So Melville won a point.

The next morning we ate before daylight. Then one of my bed partners harnessed an old horse, hooked him to a buggy and drove me to McLeod. I doubt we exchanged a dozen words the whole trip, and I wonder who he was and if he still lives up the Boulder. I never saw him again as a kid, that is. Might be I've known him for years as a man. But, to the best of my knowledge, the only Maddox I ever heard of was Pearl—and I damn sure could never have picked *him* or *her,* as the case might be, out of a crowd.

I caught the stage out of McLeod and spent another night at the Grand. By then I was getting a little blasé after all my travels. So after supper I loitered under the gallery in front of the hotel and tried to suck my toothpick with the same flair as did the other patrons of the American Eagle. Mother'd have had my hide if she'd seen and heard me.

The next morning I took the stage to Melville. Dad picked me and my riding gear up shortly after we'd gotten in.

"Good trip?" he asked. "You bet," I answered, and that was that.

Mother's reaction was a little different. I had to undress outside, bathe in a washtub and change clothes. After all, the Grand Hotel was affectionately known as the "Bedbug Rest" for a reason, and she was taking no chances.

I think the most unpleasant trip I ever made for Dad was during the summer of '24. I was almost twelve then and pretty good-sized—at least in my own opinion. But I sure got my air tested.

I'm hazy as to the whys and wherefores, but Dad took me and my rigging down to get a horse and ride him up to the Canyon Ranch. We went to Big Timber first. Dad had some business to tend to and got to talking, so it was pretty late in the afternoon when we got to the place where I was to pick up my pony. Seems to me Mr. Overfelt, a nice old Virginian and a whale of a banjo picker, was on the outfit, though whether he owned it or not I don't know. Anyhow, it was near where Swamp Creek hits Big Timber Creek.

It took some time to corral the horse, for he was running with a partner that plainly had had a shot of weed, so it was getting along by the time he was saddled. He'd been a little broncy when the cinch was tightened and snorty around anybody afoot, so Dad got up first. There was a little excitement, but after he'd thumped the colt a time or two and made him stand to get on and off of until he quit trying to spook by, Pop turned to me.

"Get aboard," he said. "You won't have much trouble with him. He's green, not mean. Better follow the county lane to Plaggemeyer Gulch and the Canyon road from there on up. I don't want you cutting across country, late as it is. Looks like you might get wet, too. You can't make it home today, but if you're lucky you may get to Vik's. If not, Plaggemeyer's. Either way, they'll put you up for the night. I'll tell them you'll be along," and he got in the car and drove off.

We followed—down to the bridge over Big Timber Creek, up the hill to the lane and turned north under the rimrocks. The youngster moved out well but kept grabbing himself every time I shifted in the saddle. So, discretion being the better part of valor, I opted for a jog rather than a lope. Wasn't long, though, before bucking off became almost the least of my worries.

Dad had mentioned getting wet, and he sure didn't have to be any hell of a prophet either. Back of, or in, the Crazies, a storm was building, the clouds so black that I could barely make out the peaks except when the lightning threw them into relief. Before long the flashes, all across the mountains, played with hardly a break, and thunder growled steadily.

I don't believe I ever knew a cowpuncher who wasn't deathly afraid of lightning, and I'd heard hair-raising stories apropos the damn stuff all my life. A high percentage were farfetched, I'm sure, but I'd seen what it could do myself. Dead cattle piled in a fence corner after a hot storm; trees and telephone poles in slivers; blue balls rolling out of a telephone on Gramp's desk; an old-time crank telephone exploding off the wall in the burst of blue flame; a hammer blown thirty feet in the air; and a man knocked down as he drove a staple into a post, trying to get the last bit of fence finished before a storm broke. Seems to me we used to have worse electrical storms as a rule than we do now. Maybe not, but in any event, I have been awful leery of them all my life, and this looked like it was going to be a booger.

It was. In the flashes I could see it break out of the canyons, and the peaks disappeared behind dark curtains. There wasn't a breath of air—just hot, muggy and foreboding. I didn't have a slicker, I doubt that the colt would have stood for one, anyhow. And there was no shelter. Just the open lane, the rims to my right and the creek bottom a quarter-mile or so below to the left, and I wouldn't have gotten near those trees for love or money! The fences on either side of us bothered me, for lightning can strike a fence some way off and follow it. That's what happened to those cattle I mentioned. But the rims stuck up a couple of hundred feet above us, and I wasn't wearing spurs—lightning hits high places, and iron draws it, so I'd always heard. Still, it was damn little consolation, for I was up on top of a horse, we were both sweating to beat the band, and lightning follows heat. Not only that but he also had an iron bit in his mouth, and I had hold of the reins to that goddam bit.

By now it was full dark, and the electrical display—what we call "dry" lightning, the wicked stuff that stays just ahead of the rain—was walking down the long flats from the Crazies. Oh, there was lightning in the storm, too, but I could see the other better. It would have been pretty, I guess, if I'd been able to appreciate it. Sometimes a crooked column would stand quivering from the clouds to the ground for a full second or more. Or perhaps a bolt would arch across from the Long Gulch to Amelong Creek, a matter of miles, and never touch down at all. And always the thunder, not growling now, but booming louder by the minute.

Suddenly, during one of the infrequent intervals of darkness I noticed a point of flame over each of my horse's ears. Steady, blue like the bottom of the flame of a Bunsen burner, and they moved back and forth as he moved his ears. I'd heard of it, for Gramp had told me of seeing it over the ends of the horns of a string of steers before a storm. Saint Elmo's fire, I think it is called—sailors know it—but it was eerie, and I cleared my dry throat and spit. There was a pop, and a blue streak down past my pony's shoulder. I don't know which of us it scared the most, but I do know that in the ensuing go 'round I couldn't have been cut loose from that colt with an axe. I wasn't about to be set afoot right then. Not by a damn sight! Just as I got things squared away again, there was a blue glare in company with an ear-shattering blast of thunder that made both of us stop and hump up, and the rain hit us like a wall.

I couldn't have gotten wetter faster if I'd fallen into a well. It came down so hard I had trouble breathing. The old colt swung his tail to it, and there we stood, even my cantleboard wet. No hail, thank God, for I couldn't have held him. I guess he didn't know about lightning for he wasn't fazed by a half-dozen pops so close that a man could smell it. I knew though, and when it finally marched away past us, my stomach relaxed some. The worst of the rain went with it, so, figuring we sure weren't getting anywhere standing still, I got us started back up the road.

We'd gone about a mile with the rain dropped to a steady drizzle, when I thought I heard something behind us. My horse did, too. He spooked, swung around and stood snorting and staring into the blackness. When the sky flared, I caught a glimpse of something, something pretty good-sized, on our track. Seemed like another flash would never come so I could see, maybe, what it was. Before one did, the colt whistled. There was an answering nicker, and a horse came up. As a flash lit things again, I could see that it was the loco we'd left back at the ranch. There was quite a lengthy reunion, but I finally managed to get started again with the poor, damn loc' trailing along at our tail. How he'd gotten there, I had no idea, but it wasn't long until I began to get edgy with him stalking along behind us.

There was a schoolhouse at the foot of the Big Timber Creek hill, and as we neared it, I began to toy with the idea of holing up for a while. Just before I got to it though, somebody called, "Joung Paul Wan Cleve?," and old John Hanson, who lived a little way off the road, appeared out of the night. "Ay bin vatching jou by de lightning. Paul

Yunior said jou vould be 'long. Coom now and eat. De missus has it ready." He damn sure didn't have to ask twice!

Old John had originally homesteaded on Otter Creek just a mile above where I live now. I guess he was cantankerous, for he couldn't get along with any of his neighbors in that part of the "Settlement" and fought steadily and bitterly with all of them. Finally, he moved down on Big Timber Creek and in short order was embroiled with everybody there. But, strangely enough, he now was as friendly as a Shepherd pup with those he'd hated so badly up on Otter Creek. I know he and Mrs. Hanson sure treated me fine that night, and I ate so much I wondered how in the hell I would be able to get ahorseback again. They wanted me to spend the night, too, but Dad hadn't said anything about Hanson's, so I figured I'd better not. So, reluctantly, but with a full stomach and a little warmer, I thanked them and started out again.

The storm had passed while I was eating. The sky was clear, the stars out, and I suddenly realized we'd lost the locoed horse during our stop. Where he went, I don't know, and I never saw or heard of him again, but it was a nice change to be rid of him. Then we turned off the Melville road into the Wormser grade and cut out of the steep hill between the cliffs and the creek. There were a number of big rocks in the road, evidently brought down from the rims by the storm. Twice I heard others come down behind me. Below, the creek was booming from the rain too and eating at the hill below the grade, so I was glad to get past it. The lane swung away from the creek, and as I rode along below the butte, I kept a wary eye on the old Wormser place below us. Only the carriage house was left, but it was supposed to be haunted. About all we needed was a damn ghost!

The lane turned north, topped the Ten Mile divide and for a while I really enjoyed the trip. The storm that had hit

us was now 'way and away northeast of us over lower Fish
Creek or maybe the Musselshell. We could see it as a big
black cloud, plain in the clear night sky on account of the
lightning. Once in a while, a crooked white bolt would
streak from the cloud to somewhere behind the horizon
and stand pulsing. Most, though, were inside the cloud,
lighting the whole of it. Reminded me of one of those paper
Japanese lanterns except that each flash lit up something
different—the constantly changing eddies, pockets, shadows and swells of the thunderhead. Seemed like the
lightning was never quiet. Before one burst had died,
somewhere in the depths another would flare, each high-
lighting a new scene. It was lovely, especially since it was so
far away I couldn't even hear the thunder.

I got so interested watching that I missed a sullen
muttering off to the west, but I finally woke up to the fact
that *all* the light wasn't coming from the display I was
watching. Sure enough, another storm was mounting back
in the Crazies. I'd shot my wad for the night when it came
to storms, so, slick gumbo or not, I kicked my pony into a
lope. I knew damn well I had no chance of making Vik's
before I got caught, so I decided I better figure on staying
at Plaggemeyer's. Dad had mentioned them, so it would be
all right with him, and when I got to their place, I turned
in.

Tying my horse to the yard fence, I headed for the house.
Keeping a jaundiced eye on a big, mean-looking dog that
was raising a hell of a racket as he sidled stiff-legged around
me like a damn wolf, I knocked. Probably I did it harder
than was really necessary, for when Mr. Plaggemeyer came
to the door, I got the impression that he wasn't real
enthusiastic about being routed out of bed in the middle of
the night. Can't say as I blamed him, either. Seems he had
thought I was long gone by, but, yah, put my horse in the
corral, and I could sleep in the granary. But be careful with
fire. So I tended my horse and slogged through the mud to

what I figured was the granary, the damn dog tailing me all the while, showing his teeth and growling. I found a chunk of board by the barn, though, so he kept his distance. I almost wished the bastard would make a run at me so I could cold cock him with it, but it just wasn't my night.

The granary was darker than the inside of a black cat, but by the lightning which was getting closer by now, I found a bunch of gunny sacks, made myself a pallet on the floor and turned in as I was with the rest of the sacks on top of me. When the storm broke, I was snug in my nest, and even though I had begun to itch something terrible from the chaff and grain dust in the sacks, I damn sure had no complaints. It was a real nice change from the last deal. So I lay there as the storm passed, the thunder grumbling away into the distance, and the last thing I remember was the sounds of mice from around me. Then Mr. Plagge-meyer woke me up for breakfast.

I had a fine meal, thanked them for everything, saddled up and was on my way by sunup. My colt had cooled down a lot—miles have a way of doing that—and we moved right along. Everything looked better by daylight; a fine, bright morning with a few puffs of clouds over the peaks. But pretty quick I noticed they were gathering and knew it'd rain again. I had it made, though, and got home in time for lunch. Dad looked up from the table and said, "Figured you'd be along about now. Have any trouble?"

"Not really," was my answer. "But," I told myself, "this is the last trip I make for you, damn you, if there's a cloud in the sky."

It wasn't, though.

16
HORSES, HOGS AND HOUSEGUESTS

WHEN WE FIRST moved to the Melville ranch the spring of '37, we didn't have any way of getting around except ahorseback or by team. Consequently, just like when I was a button twenty-odd years earlier, our kids never got to see a movie or sample a soda unless we were lucky enough to borrow Dad's pickup for a trip to Big Timber, which was seldom. That didn't mean they didn't have any excitement, though. Not by a damn sight, for Barbara used to bring them along whenever she joined me, which was whenever she had things sort of caught up around the house. The Melville country wasn't strong on babysitters, anyhow, and little kids went where their folks went— whether it was stock work or a dance—were put up where they couldn't get in the way or hurt and had a good time watching and learning.

I was breaking a filly that spring. I was doing it in the evenings for I was too damn busy to mess with a colt during the day. It was fun after supper anyhow, and she had been coming along nicely, so that evening I told Barbara, "Honey, I'll help with the dishes, and you and the kids come on up and watch how well Chula is doing," so they did. There hadn't been a corral on the place when we moved onto it, but I'd built a little square rigging by the shed with what poles I could rustle up. It wasn't much account and awful rocky, but I figured it would do until I could get to the mountains along in the fall, get out some

216

good corral material and get started on real pens. Funny, I guess I have built a good set of corrals about every twenty years during my lifetime. Built or rebuilt them, for they seem to last just about so long. Those I started that fall were my first.

Anyhow, I brought the filly up from the barn, saddled her and untracked her, all with no attendant excitement. And feeling pretty cocky, I grinned at Barbara and the kids peering between the poles and said, "Watch how well she handles," swung aboard and got stood on my head on the rocks.

I bounced to my feet, explained cheerily, "Took me by surprise," for she hadn't ever bucked before, and stepped on again. A little carefully. I'd made about one circle of the corral when she exploded again. I did a little better, but not much, and when I picked myself up, I was beginning to realize that this bay bitch could buck a little, but I caught her and got in the buggy.

I screwed down tight, ready for her to blow the plug, and she did. I took up another homestead. It was getting sort of tiresome, and when Barbara said anxiously, "I wish you wouldn't get on her again, Spike," I told her, "Goddammit, can't you give a man a little encouragement instead of just standing there like a bump on a log?"

"OK," she said, and when the filly blew again she called, "Stay with her," but the hell of it was that I heard the last two words just about the time my shoulders hit the ground.

"I'll say it quicker this time," Barbara assured me, and she did. Didn't seem to help a bit or maybe it was just plumb impossible to say it fast enough!

By now, it had damn sure dawned on me that I had myself a wolf and that either she was improving or I was getting worse, for I was lasting shorter and shorter. But she was my horse, and it was up to me to ride her, so I stayed with it.

She bucked me off either eleven or twelve times in all—I

was getting a little hazy by then and wouldn't swear to the count—before I quit her. I couldn't ride her, and both of us knew it, and while she was thriving on practice, I sure wasn't. So I decided to call it a day. As we were walking home from the barn, Barby, who was about three then, got me by the hand and looked up with a happy grin, "Gee, I hope we do that again, Dad. It was fun!"

I wasn't the only one to provide excitement around the place. Barbara sure did her share, too. Especially when it came to hogs. I don't like hogs. Probably because when I was about seven, I crawled into a pig tunnel in a straw-stack, figuring there was nobody home. I was badly mistaken, for every hog Dad owned, and he was quite a hog fancier, must have been asleep in the bowels of the stack. I spooked one, he spooked the rest; there was only one tunnel, and I was in it so I was run over by every damn pig as it stampeded for daylight. Sows, piglets, shoats, the works. Everything but the boar—he was penned up, thank God. A wonder I didn't smother, though I doubt it could have been a hell of a lot worse than the shape I was in when I finally got backed out of there.

Even so, at Dad's insistence we always kept a hog or two on our place after we moved out of the mountains. Had a "pig pail" for our garbage and fed that, skim milk and a little grain to them. I guess it's illegal now, for they can only be fed cooked garbage—or so says Washington in all its omnipotence. Be that as it may, they ate well though it did cause a little annoyance on Barbara's part when I'd remark at the table that "Jakey" or "Petunia" sure tasted good. An attendant outburst of tears from the kids followed.

Tack (who had an empathy for hogs for some reason—and never heard the last of it) acted as head hog man. One summer when he was about ten, a string of shoats some six weeks old got out of the pen so everybody went out to corral them. All but me, that is. An old pony had stove me up so I was on crutches. I just watched and tried to keep a straight face whenever anybody happened to look in my direction.

I think there were five shoats, and I'd say, conservatively, it took right at an hour apiece to get them all caught. Trouble was, that sometime or other I'd told my wife the way to get hogs to move was to say, "Boosh, boosh." So, every time she and the kids would get everything bunched and headed for the pen, she'd remember. Loudly. Each time those five shoats went in at least eight different directions; looked like a bomb had exploded in the middle of them. Finally, the idea of penning them en masse was given up, and they were hunted down or at least captured singly. Of course, the chase ranged through, among and around most of the ranch buildings, but what I couldn't see, I could follow pretty accurately by the hue and cry.

At last, all were caught but one. Two had been captured in the barn and were safely tucked in the mangers. One was ensconced in the warming box in the calving shed. Another was safely penned in my horse trailer. I'll never forget the apprehension of the last, for it took place where I could see it all.

The shoat was against a net wire fence. He was pretty run down by then, so Tack was able to ease him carefully along the fence toward where Barbara crouched under a hay rack. It was tense, tense—the shoat working slowly toward the rack, my wife like a great cat getting ready to pounce, the setting of the feet, the wriggle of the hindquarters. Everything.

When he was about three feet from the rack, the shoat's head came up. He'd seen her! But in the split second before he made up his mind, she sprang, enveloped him and clutched him under her. There was a muffled squeal, and Barbara's voice pealed triumphantly, "Gotcha, you little sonofabitch." I gave up and collapsed over the yard fence. Sobbing.

Later, when tempers were less brittle, I jumped Tack because the shoats had gotten out. He gave me a detailed account of all he'd done to fence them in. I wasn't sympathetic. They'd escaped, hadn't they?

Finally, reluctantly, he admitted that he guessed, maybe, that he couldn't keep them in. "Well," I remarked, "if you want to admit that a hog is smarter than you are, it's damn sure your privilege."

"I didn't say that," he protested.

"Hell, you're trying to keep them penned and evidently can't do it. Sure looks to me like they've got more brains than you. Wouldn't you say?"

He didn't have an answer to that. Just went up and worked like a nailer on the fence. Even spurned my offer to help, or at least to oversee. He got a good job done, and it held them. For a while, anyhow.

The pigs we had that year were those black, belted kind. I think they are called Hampshires. With short, turned-up noses, the rootingest hogs I ever saw. Why, if I was in the business of tearing up concrete or blacktop, I wouldn't mess with this high-priced machinery. I'd just get a string of Hampshires and turn them loose on the project, and they'd put a bulldozer in the shade. Only thing, a man might have trouble keeping his sanity.

Anyhow, Barbara raised a fine garden that summer. She'd hoe it in the evenings after supper and press the kids into service pulling weeds. And unless I had some real pressing business elsewhere—which I damn sure didn't dare have very often—I'd rig up the hose and irrigate it for her. But, about a month after the shoat gather, just when everything was grown and ready, one night those same hogs finally found the weak spot they'd been looking for in the pen. By morning, the garden was newly ploughed. It was my wife's last attempt at gardening, ever, for she stated emphatically—if the word does justice—that she wasn't in the hog feeding business. Nor ever would be again. And she's sure kept her word.

Speaking of gardens reminds me. At the Melville ranch we had a potato patch right out west of the house. This morning we'd just come back from church, and I was shedding my Sunday-go-to-meeting clothes. I was down to my drawers when I decided I might as well hoe a few spuds. It was a lovely day, we lived on the bottom with brush all around and it was dry enough so there weren't any deer flies, so pretty quick I was at work, wearing only my drawers and boots.

I was happy as a pup with a calf hide when, suddenly, around the bend in our road came a car. I quit the country, but as I hightailed it out of there, I could see that there were several people in it. Staring and grinning broadly. It rattled me, so as I passed Barbara on the porch I begged, "Tell 'em I'm not here!"

"No way, Buster," she came back. "Who do you *want* 'em to think is here in that getup?"

During our years on the Melville ranch we fell heir to a number of grub line riders. I've heard the practice referred to as "sundowning," but under the Crazies we called it "riding the grub line," and it was a perfectly acceptable proposition. The thing was that the ranches only kept a couple of men over for winter work and the rest were laid off after shipping and threshing were done. Usually they'd winter in town by pooling what money they had for a place to stay and groceries—the "Y" in Melville was just such a place. They might even earn a little to go with it by helping out around the livery barns and saloons and got by most of the time.

Just exactly how much it cost to winter like that was a moot question, and the story goes that several men were arguing about it, claiming this and that amount. From a couple of hundred dollars on up were mentioned as the absolute sinews of survival. Finally, an old-timer, one of a pair that had been listening quietly, spoke up, "Hell, boys, it'll cost you all you got." Before anybody could answer, his sidekick tamped it down with, "An' then some." Everybody agreed. Wholeheartedly.

So there was always a good chance, especially with a late spring, that they'd get short of money and credit—which was never too shiny anyhow. Then, so the rest could make it, a couple who had their own horses would hit the grub line. Not together, that was poor manners. Each would make his own circle.

Maybe a man had blown his wages or most of them on a drunk when he first hit town or over a card table or down on the line or it could be that he hadn't had enough to winter on to begin with. Whatever, he'd end up visiting around, too, and helping with the work at each outfit. Working for your board was about what it amounted to, so nobody minded grub line riders. Just as long as they didn't stay too damn long. Besides, they always brought along all the juiciest news tidbits.

I've known other reasons, too. A rancher friend of mine married a dude girl. A lovely person, but her husband just, by God, didn't get along with his mother-in-law. Just mention her, and he'd get a bow in his back. She'd come west and visit her daughter for a couple of weeks about every two years. Gave advance notice, so he'd rustle to beat hell getting everything done ahead of time. Then when his wife went to meet her mother, he'd saddle up, catch up a bed horse, pack a warbag and go visiting. When he'd show up at first one and then maybe six or eight other ranches, everybody knew why and accepted it as being perfectly normal. Then he'd go home again when the coast was clear. I don't see how he got away with it. I like my mother-in-law so I've never had to do that. Besides, I shudder to think what Barbara'd have done if I had. She'd have come hunting, and I'd damn sure have quit visiting. Abso-by God-lutely!

Other guys, I'm sure, rode the line just to see the

country. To find out what was on the yon side of the ridge. They'd appear, stay a while, ride off someday, and you'd never see hide nor hair of them again. One in particular sticks in my mind.

He showed up at the barn one evening, riding a good-looking paint gelding. I don't like paints, but this was an animal that showed some breeding, and his colors were grullo and white, a combination I'd never seen in a pinto before. Don't believe I'd ever seen an old boy like his rider, either. A long-geared cuss, sitting on an old, patched-up saddle, wearing a worn pair of brogans and a grin from ear to ear. "Remember me?" he asked and pushed his beat up old hat to the back of his head. So's I could get a real good look at him, I guess.

Well, hell, I had no more idea who he was than Adam's off ox. So I hemmed around, said I knew his face but couldn't call his name and so on. Finally, with a triumphant air he announced, "Shucks."

There was another expectant pause. But he'd sure lost me, so I just waited, figuring sooner or later I might get an inkling of what the hell this was all about. At last, with some chagrin he said, "Reckon you don't," and proceeded to enlighten me.

Seemed I'd given him a lift from Tucson to Cortaro two years before, when we'd had our outfit in Arizona. We'd sure had a nice chat, and he'd told me then, I was informed, that he'd drop in and see us sometime, and here he was.

Well, I didn't remember him or the incident, but I reacted like any rancher would have back then, "I be damned. Unsaddle your pony, throw him in with mine and come on over to the house."

Barbara looked a little surprised when we showed up but made the guy welcome. Not that he needed it. Short of roping him and dragging him off the place, there was no way we could have gotten rid of him. He'd come to stay, by God, and he did—for damn near a month.

The thing that had been nagging me was answered, anyhow, when I went to introduce him to my wife, I said, "I don't believe I . . .," but before I got her all out, he said, "Shucks."

There was a minute of silence. Then Barbara with a puzzled look, "What?"

"Shucks."

"Shucks?"

"Yup, Shucks. That's me."

That was his name. Whether it was his first or his last, we never found out. We did gather during his stay that he was a Missourian, but that's all. A Missourian, name of Shucks.

He was a pretty good old kid at that. He and the children hit it off right away. He was about, I'd say, on their level anyhow, and it didn't take us long to savvy that he sure as hell was missing a few buttons—for a grown person, that is. We were vindicated, too, when I asked him one day how come he wasn't in the army, since the war was on then.

"Oh," was his response, "I went in when I got m' notice, but after I'd talked to 'em for a while, they sent me to some sort o' doctor. He asked me a string of th' damndest fool questions I ever run into, an' it come to where they told me they reckoned they didn't need me. Said sumpin' 'bout mental comtipence, whatever that is."

He was a talker from away back! That was all right, seeing as we didn't own even a radio in those days, and I really think he believed all the gaudy tales he told. It's a cinch, though, that a lot were lies, for he couldn't have done everything he said he had if he'd been Methuselah's twin. He sure enjoyed telling them, we enjoyed listening to them, sort of, and the kids ate them up. He was handy with a jackknife, too, and made them all sorts of stuff. Only thing, he must have been dropped in a sure-enough poor grass year and hadn't gotten filled up since, because, my God, how he ate! Didn't seem to have any druthers—just so long as it was eatable. We weren't living too high off the hog ourselves in those days, and it sort of got to Barbara a little. Me, I just did my damndest to make sure he earned what he ate, and I think we broke about even. Maybe.

Two men can accomplish better than twice what one man can; say about what two men and a sizeable boy can working alone. Well, it would have taken two sizeable boys, for Shucks was a working booger. He sure must have come off a farm back home, though he never said, for he savvied that sort of work and when he built to something, he'd just bow his neck and never look up. Kept me rustling to keep up with him, so we got caught up around the place. More than caught up, we got so far ahead, we got things done that didn't really need doing. Then, that is. The drawback was that the harder he worked, the more he ate. Besides, I found myself doing the same. Then, one morning he told us, "Reckon I'll head on down th' road. Sure enjoyed th' visit with you folks, an' I might show up again sometime. Never know."

"Hell, don't rush off," I told him, and the kids raised such a ruckus, it looked for a while like he might change his mind. But no, a half-hour later with the paint fat and curried slick, his possible sack and henskin sleeping bag tied back of the cantleboard of his old A fork, he stopped at the house. "So long an' take care," he told us and rode off. We lived down in the brush, so we didn't see which direction he headed. Nor did we then or ever hear anything of him. Just like he came out of nowhere, he went somewhere. But right today, I wouldn't be all that surprised if he appeared. Startled, sure, but not surprised. Hell, he warned us he might visit us again sometime, and he's liable to do just that. If and when that day comes, damn if we won't be about halfway glad to see him. Shucks, yes.

Another individual we drew when we lived at Melville comes to mind. Technically, I suppose he couldn't be said to have been riding the grub line, for he was afoot. Nor was he a native, though he aspired to be. He just happened by and ended up with us—sort of like a stray kitten.

It was along in early June, and we were in the middle of one of those four-day spring rains ranchers pray for. Usually without much luck. I was up at the saloon, which was crowded with ranchers and their crews. Couldn't do much work, wet as it was, so we were celebrating the moisture. We get damn little of it in Montana—when we need it. It was well into the afternoon, and by that time a goodly amount of Hanson's stock had been taken care of. It had been a motley crowd to begin with, so things had advanced to the festive stage. Not the fighting stage; everybody was too happy about the rain. Suddenly, the door opened, and this young fellow peered in. What he saw must have made him take a long look at his hole card, for he stood there sort of hesitantly until somebody told him, "Well, come in or stay out. But shut the damn door."

He stepped in, and we all earmarked him as not belonging—the runty canvas hat, short blue slicker, muddy dress oxfords, and he was packing a guitar case, to boot. But Melville's hospitable so, "How about a snort, pardner?" someone offered.

The stranger looked bewildered by the invitation. I doubt he savvied it was to him even and moved over to stand close to the stove. After he had steamed a while, he asked, "Could any of youse guys tell me d' quickest way to get to Two Dot?"

His accent got immediate attention. "Sure," answered Sandy Harper from the bar. "You afoot or ahorseback?"

"What?"

"Are you ridin' a horse or walkin'?"

"No. Don't have a horse."

"Well," said Sandy, "that bein' so, I 'magine th' quickest way to get to Two Dot for you'd be by runnin'."

The poor devil looked so puzzled by the guffaws that followed and so just generally woebegone that I walked over and inquired, "How come you're headed for Two Dot?"

"I like de name," he confided.

"Got a job waiting? You look pretty gant."

"No. What's dis gant?"

"Hungry."

"Jees, yes!"

Maybe it was the booze I'd had, but I felt sorry for him. "Maybe I could use you. How about it?"

"Jees, yes," and that's how we fell heir to an Italian cowboy singer, would-be class, from Brooklyn.

I don't remember his last name. Probably couldn't spell it even if I did, though I wrote it on a check or two that spring. But his first name was Matt, and he was a damn nice Italian boy. Stout, eager, polite and clean. The last was important, too, for we didn't have a bunkhouse. Didn't need one, for in the summer the hay crew lived in a couple of old buildings up town, and the rest of the time I got by alone. So most of the year whoever dropped in stayed in the house with us. As a rule, grub line riders were all right— clean, only just hard up—though I did have an old sheep wagon down near the house where I could put a man if he was plumb dirty or I suspected he had bedbugs or seam squirrels.

At first, we did have trouble translating what Matt said, but we either got used to his Brooklyn language, I won't call it English, or he got used to ours. So we were able to communicate pretty well—except when he got excited. Then it was just a question of backing off and letting him cool down. When he had, we'd take another shot at it.

He sure wanted to be a Westerner. On top of that, he had plans to be a Western singer. Actually, he had a good voice, and he did a better than fair job of playing a guitar. He had his nom de plume, so to speak, all picked out, too—"Montana Matt." So we were regaled each night with music, which was fine. The kids liked it, and so did we, for a while, anyhow. Trouble was that his repertoire was limited. Damn limited. "Cool Water"—it came out "wada"—was his favorite, and to this day I can't hear it without seeing Matt whanging on his guitar and singing lugubriously. Somehow a Western song, rendered with a thick Brooklyn accent, seems off kilter, to say the least. Somehow they don't jibe; like hunting grizzly with a twenty-two, sort of.

When it came to ranchwork for him, I was stumped for a while. He could spot a horse from a cow all right, but that's about all. He did have an abortive try at an old gentle team, but inside five minutes it was about an even break who was the most completely confused, Matt or the team —or, for that matter, me. So I gave up the idea of having him around livestock of any, by God, kind. Then I happened to remember that I had some fencing I hadn't had time to get to, and figuring he'd be hard put to get in much of a jackpot with a shovel, bar, stretcher, hammer and staples, I pointed him at it. Of course, I had to demonstrate everything, but he took hold right away. Did exactly as I had shown him, and he built a fine fence. It's still there right today in good shape. Hard work, without animals, he could handle.

When he finished along before haying, I couldn't use him anymore. He had a fair stake, though, and a good pair of shoes in place of those oxfords. So he cased up his guitar, told us, "Jees, t'anks," and left. I don't know whatever became of him, but for quite a while I kept an eye and ear out for a "Montana Matt" in country music. Sort of hoped he'd make it, even though he wasn't too shiny. At that, he was better than most of what they call music nowadays. Besides, he was a good kid.

=17=
BRASS MONKEYS BEWARE!

I'VE GOT A FEW pet peeves. One is the fact that there isn't a law barring lawyers from running for Congress or state legislatures. If there were, though, the foxy boogers would probably find a hole in it somewhere, so what the hell. I've sort of given up except to daydream a little.

The other two grouches I nurture have to do, first, with people who write songs like "Let It Snow, Let It Snow, Let It Snow," or "I'm Dreaming of a White Christmas." But they only write them once, and damn few of them run cows so I guess I can't really blame them, though I sure don't agree with them. What sort of a knothead would write songs in praise of snow? Beats me. My big peeve, though, is weather forecasters. And it's a dandy.

I suppose they were normal little kids to begin with, but they damn sure didn't stay that way. They're like the two old guys that got into an argument. Finally, one called the other a sonofabitch. Naturally the recipient of the title got pretty warlike, but about then the man he was fixing to whip told him, "Whoa, now. I'm not runnin' down your pedigree. Don't mean it that way at all. You're a *self-made* sonofabitch."

I never heard the end to the story, but, far as I'm concerned, weather prognosticators just about fit that category—especially if they are skiers, too! Hell, we'll have had snow, tail-deep to a tall Indian, for the past month or two. Probably a sorry, half-hearted chinook has put a crust

229

on it so nothing can move when there's a wind, the haystacks are going down awfully fast and green grass is away down the road. A man gets to lying awake nights, humped up like some old shelly cow.

About then one of these birds will come on the screen, smirking toothily, and announce cheerily, like he'd won a big jackpot: "Here's good news! Heavy snow is expected tonight and possibly tomorrow and the next day. Just what we skiers want." And if it happens to be pretty brisk, he'll get so wrought up about the "wind chill factor" that he plumb forgets to say what the temperature actually is. I wish he'd had to stay with us ranchers forty-odd years ago. There'd have been damn little grinning!

I've heard it said that in Montana we get one month that's too damn early in the spring; one full month, if we're lucky, of summer; another month that's too damn late in the fall; and winter for the other nine. When we get winter in Montana, sure-enough winter, there's nothing between us and the North Pole but a few barbed wire fences. And most of them will have a strand or two down or broken, to boot. Most ranchers about halfway agree with what the old cattleman told Rev. Vanderbeek, the Melville preacher. I have no idea as to his persuasion, but he was a hell's fire man from away back—a great hand for stamping, too, until the time he woke up a skunk that was underneath the pulpit. Church let out early that day. He must have figured the Melville country needed salvation pretty badly—and he might have had something of a point—and was busy as a pup in a briar patch saving people. Until the old fellow I mentioned told him, "Why don't you just give up, Reverend, an' throw me 'mongst the cutbacks? After seventy-five years in Montana, hell don't faze me a damn bit. Sounds comfortin', almost, for she's bound to be warm."

Gramp always claimed that the best way he knew of shortening a winter was to give a ninety-day note in November. Maybe so, but back in the twenties, thirties and forties his recipe misfired or something. The thing was that the cold then didn't last just for a day or two—or a week. It went on and on. In '35 and '36 the temperature didn't get above twenty below for better than two months. During that time it had a habit of dropping to thirty, forty, fifty and one night, sixty-eight below. I remember an evening, I believe it was in February of '33. Hell, it had hit right at forty-five below before Christmas in '32 and hadn't really warmed up since. Anyhow, we stepped out of the house to go milk—a surefire way to warm your hands and be damned to any sensibilities as to the cow's feelings. It seemed so nice, after what we'd been having, that Dad and I shed our coats and went on up to the barn in our heavy wool shirts. Gramp arrived shortly, all bundled up, and Dad remarked on it. "Damn right," was Gramp's response. "I went back to see how much it'd warmed up, and, by God, it said twenty-two below." Shows how a man got used to it. Sort of.

I've heard people say, "We didn't know any better." Not so. There was no other way to do our work. It was a question, simply, of a man and his horse or a man and his team. That's all we had to work with—horses. No heated cabs or four-wheel drive riggings, it was hand and horse-power, all day on a hayrack behind a team or ahorseback. If you did have a tractor, it was damn near a dead immortal cinch that you couldn't start it when it was real cold. That's where the power we used had it over machines—they'd always start. Besides, as Gramp said, "Raise what you feed and feed what you raise." So, if there'd been an OPEC then, we couldn't have cared less.

There's one thing, though, that a man who used horses in cold weather never forgot, if he thought anything of them, that is: never bridle an old pony with a cold bit. His lips and tongue will freeze to it, and the inside of his mouth will get torn up. Blow on it until it frosts or dip it in the water trough if it's open. The best of all is to take the bridles into

the house the night before. I've seen many a bunkhouse
with bridles hung along the wall back of the stove for the
night. The guys were taking care of their horses, for we
owed them a lot. Could be our lives, sometimes. Most
people don't realize that when the weather was good there
was very little worry about livestock, but when things were
really salty, why that's when a man had to get out and take
care of his animals—if he was a good stockman. The work
had to be done, so we bowed our necks and did it. Still, in
spite of how damn miserable it was most of the time, a man
more or less got sort of used to the conditions. Put it this
way: I'm glad I had to do it; but I'd sure as hell hate to do it
now.

We dressed to stay warm. There was none of the
lightweight clothing we have now. Matter of fact, I wonder
if it would have done the job. To me warmth and weight
are just about the same thing. I'm not as bad as the old
cowpuncher whose grandchildren had given him an electric
blanket. "Hell, Spike," he complained, "th' thing's too
damn light. I druther sleep under a set of harness. I want
some heft to stay warm." Maybe he hadn't turned it on,
maybe he never did, for he felt the same way about "th'
damn, henskin outfit" when I asked about it several
months later.

We wore long johns—heavy ones—wool shirts, Pendleton
pants—if a man could afford them, for they cost around
twenty dollars—if not, a couple of pair of Levis would do.
But we favored wool. It was warm, and it soaked up the
sweat, for surprisingly a man could sure work one up at
times, especially pitching hay or making a run at some
bunch quitters. We favored wool socks pulled over light
silk ones, and most people wore felt shoes, German socks or
moccasins under cloth overshoes. Once in a while you'd run
into somebody that just wore boots and overshoes. I don't
know how they stood it. They were either ungodly tough or
damn fools. Same way with these pictures I see of people

wearing hats in winter. We wore Scotch caps, or better yet, if you had the price, muskrat skin caps. Sure, a man might get caught in an early fall or late spring storm when he was using a hat, but the only man I ever ran into that always wore one, no matter what the weather, was Phil Spear. If it got rough, he'd tie his bandana over the hat, over his ears and down under his chin. Happy as a bug in a rug.

Usually we wore big silk bandanas wrapped twice around the neck. Big ones, for they came in handy for all sorts of things. Those fine old thigh-length sheepskin coats were just the ticket, or a sourdough coat; damn these short, tight jackets that work your shirttail out of your pants and leave nothing between your lower back and all outdoors but your drawers. To top it off, we used chaps, Angoras almost exclusively. They were warm, stayed limber a lot better than plain leather shotguns did. Batwings weren't very popular in winter. They didn't give much protection when the wind was behind you, and the wings had a nasty habit of slapping an old pony's shoulder when there was a sudden gust, which made a lot of riders look at their hole cards. We used mitts, though they weren't a very shiny proposition when it came to handling reins or lines. There's no glove that I've ever run into either that will keep a man's hands warm when it's twenty-five below and more, and he's pitching hay. And we sure didn't have these modern battery-operated outfits back then.

Same with feet. Those German socks got almost as thick as felt after they'd been worn a while. Washed or not. If you didn't have any, a pair or two of those home-knitted wool socks the ladies in the Settlement made worked just as well. I favored boot overshoes over regular ones. I wadded paper into their heels tightly enough so they stood up some, which made it easier to hold a stirrup or quit one, or anyhow not get a foot through it if a jackpot came along. Moccasins were bad if you wore them on the outside, and it's a cinch that the Indians and mountain men must have wrapped them with some sort of roughlock. I tried them only once. When the leather gets cold, it's like being on skates, but without blades. My feet were through the stirrups most of the time, damn near up to my knees. When I got off to walk a little and warm up, every pitch I hit, either up or down, found me sprawled on my stomach or on my back, and my pony got pretty uneasy about just what the hell I was doing.

The best coat I ever saw was the old-time, honest-to-God sheepskin. Heavy twill, or maybe it was light canvas, lined with real woolskin. They had a fur collar that came up around your ears and a frog to fasten it with so you could tuck your chin down behind it. The buttons were like pieces of willow—easy to repair if you lost one—and their frogs were big and handy. Buttons and buttonholes can be damn important when a man's hands are cold. They came in two styles. One reached almost to the ground and teamsters, especially freighters, sure favored them. Took the place of the buffalo robe coats of the early days. Riders liked the shorter style. Lord, but they were good coats! I've been trying to get one for a long time, but no luck. Maybe they are too heavy for the modern taste. Anyhow, we sure swore by them.

I forgot to mention that we usually used galluses, sometimes with Pendleton or California britches and especially when a man had on a couple of pair of Levis. Heavy pants stayed put better and with Levis, if the outer pair were held up where they belonged, they kept the inner ones up, too. Then there was another proposition we used ahorseback. The gullet of a saddle picks up air and sights it in on exactly where a pair of chaps doesn't give a damn bit of protection. The cantleboard takes care of what's left uncovered behind, but there's sure nothing to break the wind in front. It got damn chilly when a man was headed into the wind, whether it was a drift of forty below air or a hard ground blizzard. So, if things were plumb bad, we'd fasten a heavy silk bandana to the back of our pants, run it through the crotch and fasten it to the front, on top of our drawers. I've seen a strip of muskrat skin used, fur in. Or jackrabbit hide. It sure worked, though it did have its drawbacks. I might add that we never put the rigging on, so to speak, in the field. Not by a damn sight! It was added as we dressed.

Speaking of jackrabbit hide, Ed Bannion, an old Klondike man, showed me a trick he'd picked up in the gold fields to keep a man's face from freezing. Take a piece of the raw hide about the size of a dime, moisten the inside and stick it to the end of your nose. Do the same with each cheekbone and, with one about like a two-bit piece, on the point of your chin. It sure works. The cold must run out of your face into the rabbit fur, for after you've been out a while the hairs frost up to beat the band—like scrub trees at timberline after a bad snowstorm. I quit using the deal because I was getting cross-eyed from trying to see around the fur on my cheekbones and nose after they got loaded. Besides, if there was a lot of wind, they tended to blow loose after they got really frosted. Still, I never froze my face when I used them.

Naturally a man didn't get into his outfit until he was ready to go. Otherwise he'd have worked up a sweat that his clothes couldn't have absorbed. We weren't very romantic looking—a far cry from TV shows or Marlboro ads—and all the clothes made us clumsy afoot and unhandy getting on or off a horse. Once aboard we did fine, but damn I hated to get down unless it was absolutely necessary because it meant that I had to go through the work of getting back aboard again. Even with a gentle horse. Oh, I've known men that didn't give a damn whether they had a bronc or not, but by God, I wanted a good, steady animal for winter riding. Might be he'd turn over on a slick sidehill, maybe even break a man's leg. Even so, the odds were that if he quit you, he'd go home, not run off with a bunch of range horses. Somebody can come look for a man if his empty pony shows up. They sure won't if he doesn't. Until maybe too late, that is, for a cripple isn't likely to last too long immobile at forty below. This thing of getting on and off, plus a few other items, brings me to something every man who ever had to make those bitter winter rides will remember vividly—and with damn little nostalgia. I'd bet my last dollar on it, for it was a dyed-in-the-wool fact.

They say time and tide wait for no man. There's something else that damn sure won't wait for a man, either, particularly when he's cold. Kind of puts me in mind of the story of the two little kids, a boy and a girl, playing together at a picnic. The urge hit him, and, being just a little guy, he naturally unbuttoned and let 'er rip. The girl watched with great interest and then remarked, somewhat enviously, "My, that's a handy gadget to take along on a picnic."

Maybe. Ask any old boy who had to do much riding back in those wicked winters, and my bet is that he wouldn't agree with her wholeheartedly. Drawers, shirttails, maybe even a muskrat hide or a bandana to cope with, and then finally those damn fly buttons—two sets if you had Levis, with their tight buttonholes—to struggle to manipulate with stiff, cold fingers. There weren't any zippers in those days. It was one of the worst features of winter riding. Chaps didn't do a thing to help alleviate the problem, either.

Unbuttoning was bad. Getting buttoned again was 'way worse. You couldn't see what you were doing, your hands had gotten colder and your fingers cramped and numb. Make no mistake, it was tough! If a guy had a partner with him, it was lots easier. You could kneel down where you could see what you were doing and button up the other fellow. Then he'd do the same for you. I've done that for Dad more than once, and vice versa when we were riding together in real salty weather.

Sometimes, though, gaudier methods were used. Charley Johnston, an old-time hand from away, 'way back told me about when he was riding for a horse outfit in the Garvin Basin years ago. They had some green-eyed boogers as saddle stock, and Charley said Jack Coates had a corner on the worst of them. Jack never gave a whoop in hell what sort of horse he rode, but in his string was one that was plumb wicked to get on. Or off—especially when a man was all bundled up. So Jack just turned things loose over the wolf's shoulder whenever the urge to go got too bad. According to Charley, the old pony invariably blew up when he got the treatment, which conjures up an interesting picture. I never heard anything about the buttoning up afterwards. Maybe Jack never bothered with that detail, for he was a wild son of a gun. Anyhow, this not-so-small feature of the romance of cowpunching was pretty beneficial, at that. We put in winters strengthening our will power. Or maybe I should say self-control.

Around here, the high plains east of the mountains, winter means wind. That's what makes it a good stock country—our grass blows pretty well clear. It doesn't always work because it is liable, all too often, to blow like the devil when it doesn't do a damn bit of good and when we need it, sull up and piddle around with sorry breezes that just put a nice crust on the snow, especially if it turns a little warm. All in all, though, we always seem to have a wind of some sort. Come spring at first we have a hell of a time standing up straight, for leaning into it, ahead, backwards or sideways gets to be a habit that's hard to break. Oh, we have wind gauges, just like civilized folks. A simple type that has been developed for our particular conditions. Just set a heavy crowbar upright in a block of concrete that is well buried in the ground, weld a D-ring to the top of the bar, preferably on the east side and fasten a good, stout log chain to the D. Be sure the chain is long enough so six inches of it rests on the ground. When the chain starts to clank, there's a breeze. When it stands out so its end is off the ground, there's a wind starting. When the chain starts to pop, it's blowing. But when links begin to snap off, then, by God, we've got a wind! Very few of the gauges make it through the winter without repairs—a foot or so added to the chain to replace what has worn off. Least that's what we claim in the Melville country.

The wind is mainly from somewhere in the west. A southwest wind can be harder than all hell but usually isn't real cold. Matter of fact, that's where our chinooks come from. It can blow the hair off a dog from the west and northwest, and those winds can be *cold*. Usually are. But deliver me from a northeast wind! They never are as hard as the others—unless it's a blizzard—but come low, bitter, steady, keening across the snow carrying tiny, icy crystals that feel like needles when they hit bare hide. When a man put in a day with a northeast drift of air, he damn sure earned his wages.

Riley Doore and I spent just such a day, and I remember it particularly over lots of other rank days because it was just too rough for hungry cattle to come to their feed. We drove to the Briley place on the Sweet Grass in a little sleigh, then hooked to a bob with a rack and crossed the flat to the Butte ranch—about five miles from home, all of it into a northeast breeze. We made four trips across that open flat and fed two loads along the creek, but when we got back with our third load, not a damn cow had quit the shelter of the brush to go to the hay. So Riley hung the lines on the standard and fed, while I crossed the creek, got a saddle horse and came back to move the cattle onto the feed. I fought them for maybe an hour, but it was just so wicked that the cattle wouldn't leave the brush, hay or not. I didn't have chaps, and when we got home, I found I had frostbitten my knees badly where my pants had been stretched tight over them. Still, as far as looks went, the day could have been called nearly pretty—a silvery blue color over everything, with the Butte and the Crazies soft and fuzzy through the haze—but that damn breeze whispering nastily across the drifts. . . .

The haze is frost haze, and we ranchers here know it well. Too well. I don't really savvy what it is or what causes it, but I damn sure know what it feels like, for it seldom shows up unless the temperature is down to twenty below or worse. It isn't a cloud or fog. It's more of a mist, maybe. A dry mist. With an overcast it seems to cover everything, though I don't believe I've ever seen it when there was much wind. On a clear morning it lies along the foot of the Crazies and across the flats. A silvery blue, and it's only when the sun starts to touch the top of the peaks that a person can see that they are not as starkly edged as they should be. They're softened by the haze. As the day goes on, this haze backs into the canyons, filling their heads with a cloud of deep blue. It lurks there until towards sundown. Then it creeps out, thicker than before—a dark, frosty blue which blurs the outline of the mountains, somehow adding immensely to their size. If it's cold enough, perhaps sun dogs will flank the sun, burning sullenly until all three set. If it has been overcast, often there will be a touch of pink or gold with the silver. Lovely, if you aren't too damn cold to appreciate it. Charley Russell is the only man I know of who captured it. Maybe because he *knew* it, just like everything else he painted.

Funny thing about the wind. Though it can, and very often does, go on day and night for a week or more—which has led to divorces, wife beatings and an occasional killing —if it blows all night, seems as though it will taper off and quit about sunup. If it blows all day, it's likely to ease off towards sundown. Often on the Butte ranch, which was a windy son of a gun, we couldn't feed till almost dark. No use putting out hay that would end up in Dakota. Reminds me of Mr. Harry Hart. The saying was that if there was a hard frost, he'd start feeding, and, by Jove, he saw to it that it was in the morning, on time. One really windy winter he complained to Dad, "By Gad, Paul, I feah my bally cattle are losing weight every time they are fed. The blightahs spend their time chasing wisps of hay about the field at an all-out run. Else they do, they cawn't catch it."

What winter riding is like depends entirely on what sort of a day it is. A ground blizzard, for instance. You don't *go out* in one, you get *caught* in one. You'll be at work, here she comes and whatever you are doing with stock usually comes to a screeching stop—unless you happen to be moving them with the wind. If it's in your face, you might as well quit. I was halfway across the Melville flat, heading for the brush along Sweet Grass with a bunch of heavy cows, when a ground blizzard started. Right in my face. It wasn't far to the creek, and I thought I could hold them till we got there. Inside a minute I had trouble seeing my pony's ears, but I crowded up on the tails of the cattle and kept them moving, especially two old ladies that were in the drag, for once in a while I could catch a glimpse of *them*. Finally, we hit the creek bottom, and I made my count. Two head. The rest had drifted off and back past me, and I hadn't seen hair or hide of them as they did. I made another trip the next day when the wind had quit.

One time Dad and I made a horse gather on the Butte ranch of unbroken stuff four and five years old. We had them bunched and headed for the corrals when a ground blizzard howled down from the American Fork pass. The horses immediately drifted into a deep coulee. We went with them and got down in what shelter our saddle horses and the others offered. Most ground blizzards are only moving what snow is on the ground, but there were some soft flakes in this. Dad told me, "I think maybe this started as a squall. We'll wait and see if it lets up." So we humped up with those horses for an hour or two.

Those youngsters normally were a pretty goosey outfit, and I was surprised how quiet they were as we waited. Both of us got down so we could stamp our feet warm, but the colts didn't fuss about it. Even, after it slacked off a little, when Dad led his horse up the coulee through the bunch, nobody offered to kick or spook, just moved out of his way. I've noticed that most stock tends to gentle down a little in real rank weather.

Anyhow, he decided the wind had eased enough, headed up the draw at a walk and I started the outfit after him. We finally topped out at the head, pretty high on the Butte. Up there the wind had dropped to a stiff breeze, and the moving snow was only about stirrup high. A bunch of ghosts couldn't have been any whiter than we all were from the snow beaten into hair and clothes. Looking back, there was a moving sea of white from the Crazies to as far as we could see. Only the tips of the Cayuse Hills close by to the east, the nipples of the Squaw Tits 'way and away southeast toward the Yellowstone, Wheeler Butte ten miles to the south, and the dark timber on the top of Wolf Butte stuck up like islands in an expanse of moving snow. And if you watched too long, all these seemed to be moving against it. It was a sight! No creeks, no hills, no anything. Just a tide of white surging east, and the few tops stuck through it looked almost pitiful. It was one of the few times I've ever gotten a bird's-eye view of a ground blizzard. Mostly I've been in 'em and wishing to hell I wasn't.

I hate ground blizzards, especially from the northwest, for they can be nearly the most unbearable of all winter riding. Cold to begin with and that cold is compounded by hard wind and cutting, lashing snow crystals. It's hard to breathe without choking on the damn stuff, eyes stick shut, and the cold works in to a man's marrow. About all that can be done is to hunt some shelter—timber or a creek bottom. Or give your horse his head. A horse is a lot more philosophical about weather than a human, but he doesn't like it a damn bit better. He knows where he is—you can bank on that—but won't hurry where he can't see. He'll get you there, though maybe not as fast as you'd like.

I know that the coldest I've ever gotten was in ground blizzards at zero or thereabouts. Thank God, it usually doesn't blow hard when the temperature is way down. If it does, then that's when stockmen go out of business—not that they want to, but by spring they are damn shy of stock. I've come in, after having the pleasure of bucking a ground blizzard, unable to do more than mumble, holding onto my horse when I got off him so I wouldn't fall and stumbling stiffly around while I unsaddled and took care of him. When I'd finally lurched down to the house, I'd strip down to my normal clothes—or less—and back up to a roaring fire as close as I could stand it. Which was damn close. Invariably, when I was hotter than Billy hell on the outside, all of a sudden I'd get a fit of the shivers and get cold all over again. Seemed like it worked out from deep inside, and it took a number of go-rounds until it finally quit coming. Other features of warming were swollen lips, fingers that looked like hot dogs—and were just about as stiff—cheekbones that a man had trouble looking out over, ear lobes that felt as though they were resting on your shoulders and swollen toes that drove a guy crazy with their itching. But at least, by God, a man was warm. All the rest a little unpleasant, that's all.

I learned something else about the effects of cold when I made a trip to Melville for the mail one real chilly January day years ago. I got our mail and Andersons', for I'd go right through their place, sacked it up back of the cantle and dropped into McQuillan's, figuring a shot of moon-shine would help warm the trip home. Well, one thing led to another, so I had several. Then I got ahorseback and went home, leaving their mail with one of the Anderson boys who was coming in from the barn. Time I got to our place, I had begun to wonder whether old Jim hadn't taken to cutting his whiskey pretty heavy, for from the way I felt I might just as well have been drinking water. I found out different after I got in from the barn and sat down by a nice hot fire while Gramp was getting supper. By the time he called me to come eat, I was so drunk I had trouble making it to the table. Damn, but I was glad the family was gone, and we were batching. Since then I've been leery of mixing booze and cold weather—unless I was inside and figuring on staying there. Could get a man in a bad jackpot.

I've read of "whiteouts," but I don't believe I've ever seen one. We do have something along the same lines, and I'm sure every rancher up here has experienced it. Sort of interesting really, when I think about it. It'll be a day when it's snowing some. Tiny flakes that make the air misty, but the visibility is a couple of miles, so you saddle up and start out. Trouble is that those days seem to thicken as time goes on, and the distance a man can see gets less and less, until, about the time he gets back up into the hills, he is riding in solid white. Might as well be packed in cotton, no sky, no horizon, nothing except for the snow on the ground right close and that's hard to make out unless there's some sort of movement on it. Maybe a low wind will be moving it—little rivers of dry snow hurrying out of nowhere, rustling past on the crust and disappearing into nowhere again. Surprising the changing shades of white in them.

It's hard on a horse, for sometimes the snow is crusted enough to hold him until that last foot comes up, then all four break through. Tough for an old pony to buck that sort of thing. It pays to know the lay of the land, if you can figure out where you've come from since you last could see anything, but mainly you have to depend on your saddle horse. Don't spur him where he doesn't want to go. Coulees are bad medicine. In clear weather a man can see to ride around their heads, even though he may have to travel a mile or more to gain a hundred yards, because if a horse gets into one, you're in trouble. Not as bad as he is, though. If the snow is soft, which is damn seldom the case, a man may be able to tramp it enough so his pony can turn around and get back out. If it's badly crusted, you can't tramp it, and the animal is likely to fight until he gets down; and when a horse does that, he's liable to give up. So a shovel is called for or another horse to snake him out—both of which are in short supply when a man is alone. And if you do get him out, it'll probably be with something broken or dislocated. If you can get on a ridge it helps, for ridges usually have been blown fairly clear. The drawback is that ridges end, and there'll always be a drift to get through getting off them. A man and his saddlehorse, too, sure get to watching for a sagebrush or a few spears of grass—anything to show where the drift is the shallowest.

Funny how those days play hell with depth perception, especially if they clear a little once in a while. I've been traveling along blindly, when suddenly a tree or two, or some brush, will show up where there shouldn't be any. If I'm where I think I should be, that is. Then, after we take a few more steps, things fall into place, and I realize it's just a couple of weeds or a small bush sticking through the snow a few yards away. Or, after a few go-rounds like that, what I take for weeds suddenly turn out to be cottonwoods a half-mile or so away below the ridge, and then I realize the day has thinned. It's a damn funny feeling, but when they are actually trees, a man can get his bearings again. A day like that can be chilly, but what's worse, it's ungodly lonesome. A horse doesn't just carry you, he's damn good company.

Something else I've seen in the middle of winter which sure stumps me: lightning. Usually it happens on a reasonably cloudy day, though not always. I came in with a load of hay and was unhooking in the shed lot. It was snowing hard with icy little flakes and a driving wind. Just as I got the last tug loose, a blue glare lit the snow, and thunder cracked like a charge of powder. I was so surprised I jumped for the lines too late and lost my team. Luckily, we were in a big stout corral, and the team loose from the bob sleigh except for the neck-yoke, which they took with them. Otherwise, there'd have been a hell of a wreck. I haven't had it happen often, but it sure can, and I've never seen it except when it was below zero. Sort of spooky— lightning and thunder in the dead of winter. We have enough of the damn stuff in the summer.

Absolutely clear days—which are very much in the minority—I always sort of enjoyed, even though they could be colder than Greenland's icy mountains. Thirty, forty, and more below zero but with no wind. A man wasn't riding for pleasure, not by a damn sight, but the work was easier when it was clear. A man could see where he was and spot the bad places—the coulees, the cutbanks, those damn wide swales that drift in so badly but look level, and the dirty, deep little draws dropping off the side hills. All of them hazards. Besides, it was a lot easier to locate stock and handle them.

Colder than all hell, sure, but a man learns to keep warm, or at least to keep from getting too cold, by movement. A steady twisting of face muscles helps a lot. Sort of gets to be a habit. I've even seen men do it after they got back into the house and were warm. An outsider, meeting one of us back in the hills would have probably figured we were weedy or something from the way we would be squidging up our faces. If it got too bad, a man could use that versatile bandana to tie over his face, too.

A rider alone had to bank on the feel to tell when a cheek or nose was frozen, but if he had a partner, one or the other would say, "Better rub your cheek, it's started to turn white," and it could be thawed, even if it meant taking off a mitt for a little while. Hands aren't too rank, for you can hang your reins on the saddle horn and swing your arms across your chest until they warm up. Feet can be bad, and it's a question of wriggling your toes. When they get numb, you get down and walk a while, holding onto a stirrup. That's when a gentle horse comes in handy, though I believe in using heavy reins in the winter and tying the near one to my wrist when I'm afoot that way. Just in case I fall down or something and spook him, for it's a long way home if he gets loose. Especially carrying all those clothes.

Cold as it was, those clear days were lovely—a world of every shade of blue imaginable, gold, silver and white. Blue shadows were in every pocket and dip in the white snow. The timber on the flanks of the Crazies were almost purple with the shadowed walls of the canyons a deep indigo. The bull pines on the rims, the brush along the creeks, the buttes, each showing some variation of blue, and all touched with silver from the frost crystals in the air. The further away, the more silvery, until the Bearteeth, 'way to the southeast looked like they were etched in sterling. And it all kept changing as the shadows shifted. The sky above the horizon was a pale azure, gradually deepening toward the zenith, though a winter sky never seems to show the warm color it has in summer. High to the south the sun, a ring of gold connecting the four sundogs that flanked it above, below and on each side. Very likely there would be a shimmering cross through the sun in the center of the circle, from top to bottom and side to side. If it were cold enough, I've seen another set of sundogs outside the first, above and to the sides, with the horizon blocking a view of the fourth. They would be fainter and what should have been the circle connecting them looked like a dim rainbow 'way and away. I never tired of looking, just looking, on a day like that, even when I was colder than Billy-be-damned.

Another queer thing was that there were mirages. I don't know why they came about, but I've watched the hills east of Melville lift and change until they looked like rough badlands. Or the piney hills on Fish Creek, normally out of sight, would climb over the horizon, and the Cayuse Hills would sport cliffs and pinnacles. Other people have seen them too, but I never heard of anybody seeing any water, which seems to be pretty common in stories of mirages. It's understandable—even in a mirage there'd be damn little water at forty below.

As the sun swung towards the west, the silvery color would thicken until all the blues paled into it. Then, as the sun set and the sundogs faded, a last golden ray or two would mix with the silver, and the peaks turned dark with their outlines like knife edges against the last fading bit of brightness. The bitterness of those days was worth it, just to see them, to be part of them and to have the memory of how absolutely beautiful they could be.

Riding on a clear winter's night can be an experience, too. When there is new, soft snow, there is almost no sound. Hooves are muffled, and even the squeak of leather is muted. Your horse drifts across a world of white, nearly featureless and flat, and the only way a man realizes there are hills is the pitch of the animal under him or the skyline rising to block out the stars. Stars are completely different in winter. Blue-white, cold, and more of them than in summer. The sky is packed with them; bitter, sharp, like bits of shattered steel.

Usually we have the northern lights in late fall or early spring, but I've seen them a few times in the dead of winter, and they put on quite a show then. Spears of light will shoot up from the northern horizon. Up and up and then fall back into the glow along the skyline, a glow just short of a sunrise. Or, as I have seen a time or two, they may be rippling curtains of light, gently colored orange, red, blue or green, or a combination of all four, and a man can almost hear them whisper as they move. I remember vividly once when they were like a fire with wind-blown flames, the ends of which would break away to climb the vault of sky and die to the south, dimming the stars as they passed. Lovely, cold and alive. I've ridden many a mile and hour, just watching. Alone with a horse, the pulsing sky overhead, the luminous light that snow gives darkness, a

faint smudge across the white marking the timber along
the Sweet Grass, and the half-seen, half-felt loom of the
Porcupine Butte beyond. To the west the Crazies lift
pearly and unbelievably tall.

Somehow those nights gave me the inclination and the
time to think—to ponder what I ordinarily didn't bother
with. I liked those quiet rides, though it was a cinch that a
coyote or two would tune up, their voices carrying for miles
through the cold air. Just so damn happy to be alive that
they had to sing about it.

There was another sound I used to hear sometimes on
those nights. Its thrumming could be faint, and God knows
how far away. Or close at hand, filling the night. I won't
hear it again, for the wires are buried now, but I'll never,
ever forget the hum of the telephone wires in winter. It
seemed to come from everywhere. The coldest, loneliest,
most disembodied sound I've ever heard. The absolute
personification of Montana winters, as I know them.

Seems like I've gone on and on about winter, but hell, the
way I figure, I've spent right at three-quarters of my life
coping with it. Close to eight months out of every year I've
lived. I've not said anything about sure-enough blizzards.
Not that we don't get them, but under the Crazies with the
Belts and the Snowies to the north, we don't often get
storms such as those on the open country of eastern
Montana and the Dakotas. I haven't said much about
chinooks either. Might be I'll write about them both
sometime, but right now I've had too damn many memories. Plumb to my eyebrows!

One thing, though. R. B. Cunningham Grahame, "Don
Roberto" to the great horseback men of the Americas, once
wrote: "I cannot conceive of a Heaven without horses." I
agree. But if Heaven is all it's cracked up to be, somewhere
in it will be a house with a snowstorm whipping around it,
windows bright and beckoning. Inside will be warmth, a
tang of wood smoke and hot metal, with the scent of fresh

bread, spicy food and strong coffee. And those lovely, welcoming words, "Come in and warm up. I'm glad you're home."

18
THEY QUALIFIED

I SUPPOSE THE people I have lived among all my life in the Melville country were not a lot different from those in any ranch area of Montana. Montanans, particularly those east of the mountains, are a breed of their own. But to me there's a special strain of that breed around home. They have a cross-grained humor, a way of saying things, of sizing them up, coupled with resounding individualism, bone-deep toughness and a slaunchwise optimism. There's the ability to see something funny in damn near anything and the capacity to laugh at themselves. I have a very soft spot for all of them. Some may be a little rough along the edges, but they are good people.

Dad and I rode into the Billy Creek outfit years ago and were greeted by a man who was skinning a coyote on the porch. "Get down," he invited. "I'm just about to fix a bait. Come in an' eat." He straightened, closed his pocket knife, wiped his hands on his britches, walked in and started mixing a batch of biscuits in the top of the flour sack. Dad glanced over at me, grinned and said quietly, "I don't imagine a little coyote hair or blood will hurt us any." So we ate, and he sure made dandy biscuits.

Brings to mind some other biscuits I ran up against one time. It was back in '39. We'd brought up a string of Mex steers that spring and had hell keeping them on our range. There was plenty of grass, but when those cattle were filled up and happy, I guess they got curious about what was on the yon side of the far hills. Looked like they'd never seen a fence down home in Sonora, for they paid no attention to them. Maybe they mistook barbed wire for a new kind of cactus. They were used to the latter, God only knows. They didn't run through fences, they *walked* through them like they weren't there.

So we were kept busy gathering Mex steers. We got irate phone calls from homesteaders all over the country. From Mud Creek, from down in the Big Coulee, Wildcat, White Beaver, places miles away. Calls that were all about the same: "There's a bunch of your goddam steers here raisin' hell with my crops an' garden. Can't keep th' fence-crawlin' bastards out. Better get 'em, or I'm about to beef a couple." We kept a pickup with a camp outfit in the back, a trailer hooked up and ready to go. There'd be two of us as a rule. One to drive the rigging, the other to handle the cattle. If there were a lot, we might take two horses and three guys, but the steers were usually travelling boogers. Bend them the right way, and off they'd go. I don't believe they gave a damn where they went, just as long as they went. If it was too far back to our range, the truck driver'd make camp on some creek, if there was one, or anywhere that we could hold the stock overnight. He'd have supper ready when the cattle got in. We covered a lot of new country that summer. So did our horses and the steers.

Well, I was bringing some of the travelers back. On the way along in the afternoon we passed a homestead, and I stopped to see if my horse and I could get a little water. It was damn scarce in that country—on top anyhow. There was plenty about five hundred feet straight down, and this outfit had a windmill. A few little kids scattered like antelope when I rode in and disappeared. I wondered idly where they went, for it was flat as a table for at least a quarter-mile in every direction. The man wasn't home, but his wife came to the door. Sure we could have some water. Right over there in the trough.

The water was a little green around the edges, but it was wet, so my pony and I took on a load. When we'd finished, I went up to thank the lady. She'd combed her hair and swapped dresses and told me, "Come in and eat. I'm warming the stew we had for lunch, and I just put a pan of biscuits in the oven." That was nice of her.

I'd eaten 'way before daylight, and by now my stomach was sure my throat was cut, so I said that'd be real fine, hobbled my horse and went in. I think she was plumb starved for news. I wasn't too well acquainted what you'd call locally, and she didn't know much about the Melville country, so I couldn't give her any juicy bits. I was sorry about that. Didn't seem to bother her, for she talked steadily all the time I was washing up and she was putting food on the table. She was a good cook! The stew was thick and rich. I spotted it as antelope right away—hell, I did a little poaching, too. Biscuits right out of the oven, and she sure had the touch with 'em.

I was really filling up, answering her, if she gave me the chance or my mouth wasn't full, when I heard my horse whistle. I looked out the window beside me, saw what was bothering him, and found out where the kids had gone. What he was watching were their heads, poked up out of badger holes. I'd seen kids take to the timber when you rode into a place up along the mountains, but badger holes were sure a new wrinkle. Besides, I'd have been leery of rattlers.

I was well filled up when I reached for one last biscuit to sop up the gravy on my plate. I had trouble pulling the top and bottom apart. Seemed sort of stuck. I worked at it surreptitiously, and finally they gave, and damn, I was glad I hadn't drawn that biscuit sooner. In its center was a gob of hair. Long hair, twisted up like it had come out of a comb. Human hair. I swallowed a little hard but slipped the works down and into my chaps' pocket. The lady had been so busy talking that she'd missed what had gone on. I was glad of that, for she'd have felt like hell if she'd seen it. She was plumb all right; maybe a little careless, was all. I used a spoon to clean up my gravy, but can't honestly say I enjoyed it much. I drank my coffee, all but the bottom, thanked her real nicely, said my steers might have headed the wrong direction and left. I've never seen her since, don't even know who she was. But she was a generous, kind woman.

There was another female homesteader down by our lower range, but she was a different caliber than the biscuit lady. A pretty rough old blister. She had a few cows and a real sorry bull, all of which we kept finding in amongst our cattle. Finally, one day when we ran onto her shelly old cows and their peaked-assed consort for the umpteenth time, Dad told me, "Let's go over and have a talk with that old girl. I'm tired of this."

We gathered her cattle, hurried them over to her place and rode up to the house. She came out, friendly as a skim-milk calf. Dad wasted no time in telling her we'd had to move her stock off again. Of course, she professed complete surprise that they'd been on us—just as though she hadn't put them there some dark night. "I don't mind your cows so badly," Pop went on. "They don't eat much, looks like, but if I find that damn bull in with *my* cows again, by God, I'll cut him."

It never fazed the old devil a bit. "Fine," she agreed heartily. "I been plannin' on cuttin' th' sorry sonofabitch. Le'me know when you aim to do it, an' I'll bring over some Lysol an' a whetrock."

Damn seldom did I ever see Dad at a loss for words. He was then! I didn't dare look at him until we left, I wanted to laugh so badly. We cut the bull, too, next time he showed up but didn't invite the old girl to come help.

There was a bachelor down near us. Nice old fellow. I ran into him one day following along behind a handful of cows and calves. He had a couple of branding irons across his

cantle and didn't look too comfortable. I inquired where he
was headed. "I'm brandin'," was his answer. "I been
pinchin' my ass with these damn irons for better'n a hour,
but there's an empty hay corral down th' coulee a ways, an'
I can get shut o' settin' on them there."

It tickled me so much that I rode down to the corral and
helped him brand. I was in no hurry, there were only three
or four calves anyway, and he sure was appreciative. I
couldn't help but wonder, though, how he'd have managed
if I hadn't come along. But he'd have done it.

A friend of mine near Melville got to drinking pretty
heavily. "Pretty heavily" by local standards is the same as
"a hell of a lot" anywhere else, and it began to show. I met
him on the road one day, and we whoaed up to talk a little.
He looked better than he had for a long time, and I
remarked on it. "Guess it's 'cause I quit the booze," he
explained. "I got pretty badly spooked an' haven't had a
drink since."

"Sure enough," I told him. "Glad to hear it." I meant it,
too, for he was a damn good man and had just been plumb
wasting himself, to say nothing of what it was doing to his
family.

"Tell you how it come about," he went on. "I got pretty
bad, an' one day when the wife an' kids was at church I
decided to kill myself. Hell, I wasn't no good to them.
They'd be way better off without me. Well, I got down the
old 30-30, checked an' found there were three shells in it.
That ought to be aplenty for what I had in mind, so I got
ready. I was goin' to get her done real quick before I lost my
nerve, but then I got to thinkin' how bad the family'd feel if
I did th' business in the house an' blew blood an' brains all
over everything. So I went down by the Sweet Grass where
it wouldn't hurt if I messed things up."

He was quiet a minute. Then, looking me square in the
eye, "You know, I fired all three of them shells. Three
shots, an' every one was a miss." Serious as a preacher,

" 'Course they was all runnin' shots." I damn near hit him!

In the Melville country if we used the term "a character" to describe someone, you can bet your bottom dollar he sure qualified. He'd probably have been put away anywhere else. I remember one real well. He was a character—and I use that designation in the Melville sense—from away back. He lived out on the divide, a bachelor. Once in a while he'd come to town. For cartridges, mostly, or radio batteries, though sometimes he was after beans. He could order them from Sawyers in Big Timber by the hundred-pound sack, and the stage would bring them to Melville for him. For several reasons, I don't think he ate much but beans and antelope he poached, for he had an old, long barrelled 25-35 and he could sure use it. Either of his two staples makes for flatulence. Together they are downright dangerous. I mentioned the fact when I ran into him one day. "Yup," he agreed. "They damn sure blow me up. But, hell, saves on fuel."

Down where he lived it was a long trip to the mountains for firewood. He made a trip each fall with a team and running gear to get out his yearly supply, so I could see why he'd save fuel wherever possible. But I couldn't figure how what he ate had anything to do with it. So, "What do you mean?"

"Why, them long winter nights. Let a big un in bed, an' it's as good as a stick of wood on th' fire!"

I'd never thought of it from that angle. I wish I'd asked him what he did in the summer.

I missed quite a deal over in Harlowton shortly after the war. I grew up with one of the participants, and he was always picking some sort of trouble. Pretty fair with his hands, quick and catty, so he usually did all right in his battles. But this time he chose some soldier back from the war and got his clock cleaned.

Bill Donald, a neighbor over on the Sweet Grass, saw it from start to finish and told me. Seems the Harlowton boy wasn't warlike at first, but the Melvilleite pushed him a little too hard. Finally the old kid downed him, kerbooey. Bill rushed in, set him on his feet, said "Go get him," and headed him for the soldier. Who promptly decked him again. Bill got him up, said "Sic 'em," and lined him out once more. With the same result. I guess the soldier knocked the guy down six or seven times, and Donald would hurry to his rescue with words of encouragement. The last time, as Bill told it, when he ran in to get his man afoot again, the latter looked up through the blood and said in a pleading tone, "Le' me lay, goddammit. Le' me lay."

I sure wish I'd seen it, for not long before I'd told this old boy "You better be careful of these youngsters who've been in the war. They're liable to be rougher than cobs." I guess he learned a lesson, though, for I never heard of him tackling the military again. Or of his getting Donald to back him in any sort of a fight.

I always liked the McDowells. Their Dad was a fine old Texas gentleman, and Aubry and Mabry started me rodeoing. They came to every Melville show with Mabry's little bay rope horse, High Power, and usually won some money. After they moved down out of Bridger, I'd always drop in and see them when I went through. One time I stopped to see Aubry about something and found he was out in a cow camp on Cottonwood, if I remember right. I had a Jeep pickup, they told me how to get there and I drove over. Aubry was glad to see me and insisted I spend the night. Hell, there was plenty of room in his bedroll, and the water wasn't too bad if you made the coffee stout enough. I wasn't in any hurry, so I stayed.

I won't forget that night! He was camped in an old one-room house, with his bed rolled out on the floor. He got supper while I tended his horses, we talked a while and got ready to turn in. Lord, it was hot, but he had the two windows open. I came in from outside, and he cautioned me, "Shut that screen door plumb tight."

What for, I wondered. The damn thing had so many tears in the upper half that a sage hen could have negotiated it without losing a feather. The bottom half was tight, but what good would that do? I said so, and he grinned. "I got company. Listen." He walked over and stamped hard on the floor alongside the bed. Right away at least a couple of rattlers buzzed from down below. "It keeps those bastards out," he explained.

I didn't sleep real well that night. Especially when I remembered the open windows. I couldn't help but ask about it and was told, "They haven't bothered 'em any." I'd have felt a hell of a lot better if he hadn't added, "Yet."

I guess a man can get used to that sort of thing, but I didn't stay around long enough to find out. The next morning I pulled out. *After* daylight.

Snyd McDowell was another of the brothers. He was a salty bronc rider at one time, a good roper and a popping good cowboy any way you took him. He always had something to say about most everything, and it was worth listening to. He had a way of putting things, a wry sense of humor and a sharp mind that sure tickled me. I was helping him brand down on Clark's Fork one time. I wondered whether his fences were poor or his neighbor's cattle were awful breachy, for he had the damndest string of different brands amongst the cattle we were working. A whole gob of different irons in the fire, too—those of a lot of his neighbors, I guess. Anyway, whenever a calf was dragged out, Snyd would say what brand should be put on it. A calf came in that belonged to someone who had a diamond in his brand. A heated argument started between the wrestler and the man with the iron as to how the diamond should be put on. Finally Snyd spoke up and sure ended the dispute: "Hell, don't look to me like you can get a diamond on wrong." Come to think of it, I don't believe you can.

The same day one of the ropers brought in a big, stout,

early calf. Snyd rode in to heel him, but his son, Jay, told him, "Let me throw him. I need th' practice." So his dad backed off. The calf gave Jay a going over. About the third time it knocked him down and got away, Snyd, who was sitting watching with his leg cocked over the saddle horn, remarked casually, "Practice, huh? By jokes, don't know as I ever seen anybody who needed it worse."

There's another story about the man I've heard told. I don't know whether it's true, but it's the sort of thing he'd do just for the hell of it. Seems a neighbor came by, and Snyd flagged him down at the barn. Pretty quick he inquired, "Where you headed?" The neighbor said he was going elk hunting. That sounded good, so Snyd asked if he could go along. That was fine, so he climbed in. When they got to the house, he stopped to get his rifle. While he was looking for some extra shells, his wife came in and wanted to know what he was up to.

"I'm goin' up on the mountain with Buck an' see if I can get us an elk," he explained.

"But I haven't a stick of wood in the house," she protested.

"Well, hell, I ain't takin' th' axe," and away the two hunters went.

Bert Blethen was a State of Mainer, like so many early arrivals on the upper Musselshell. He was a blacksmith, a damn good one. He first settled in Martinsdale and later moved to Harlowton, where he had something to do with the Harlow-Lewistown stagecoach line until the arrival of the Jawbone Railroad. At one time he was fire chief of Harlowton, I believe. Anyhow, they tell of when a fire broke out in a barn, and Bert hurried down to its succor. A knotted rope was thrown over the peak—luckily it wasn't terribly high—and Bert started up the rope on one side and another volunteer fire fighter up the other. The other guy got to the top a little in the lead, turned loose of the rope and grabbed the ridgepole. Naturally, Bert, who was just about one handhold back of his partner, ended up flat on his back on the ground, the rope on top of him. Fortunately, he wasn't crippled, only badly shaken up, but from then on he was damn touchy about either man letting go of the rope before both had reached the peak of a roof. Don't know as I blame him; must have been a hell of a feeling when his rope let go just as he reached for the ridge.

Also, when his wife went down town to buy a dress the old fellow always went along. When she found one she liked and put it on, he'd make her do knee bends, touch her fingertips to the floor, all sorts of stretching exercises and things to make sure the seams would hold. It got so the merchants sort of hated to see the two come in, for if anything gave, he'd make her pick another and give it the same treatment. I wish I could have seen it. Barbara did a time or two when she was a little girl and said he sure made his wife give the clothes a first-class going over. Good business from Bert's standpoint, if not so shiny from the storekeeper's.

Back in the old days a man who had an eating establishment over on the Musselshell got in a supply of toothpicks. Naturally, the more cultured of his patrons had used them before, but to a lot of the local gentry they were pretty much of a new fangled proposition, but nevertheless very popular. He also made a habit of charging salesmen and other outsiders four bits for a meal, and thirty-five cents for the local people. It happened one day that a whiskey drummer paid his fifty cents, tucked a half-dozen toothpicks in his vest-pocket and left.

Behind him was an old kid from the head of the crick. As he paid his thirty-five cents he remarked, "I see now why you charge them fellers more'n us."

"Why so?" the proprietor asked idly.

"Why, hell, d'ya see all them toothpicks that man took? Me, now, I take jest one. Use 'er an' put 'er back. No damn wonder you made him pay more."

There was a man down by Big Timber who sure liked his booze. He didn't give a damn how good it was; just so long as there was plenty of it, so he favored the cheapest whiskey he could buy. One day in the liquor store a friend was surprised to see the man buy a fifth of Jim Beam. The latter cost quite a bit. His curiosity got the best of him so he inquired how come the old boy had switched—making a mental note to pay him a visit real soon. "It's th' square bottle," the drinking man explained.

"What the hell's that got to do with it?"

"If she tips over, she don't roll under th' bed. Hate to think how much time I've wasted tryin' to locate them damn round bottles that have rolled plumb over into th' corner. Man's got to get outa bed an' crawl under for 'em. Some of 'em even break. Druther pay more an' have it handy when I need it." Actually pretty sound thinking.

Three Finger Bill Davis was a hard character who lived out north of the Yellowstone on Nelson Creek. Hi Whitlock, who came from the same country, told me a story about Bill. The old devil always packed a gun, was pretty impulsive with it, and one time got into a hell of an argument with a man known as the Dutchman over at the 44 horse camp. They were by the kitchen table at the time. Finally the Dutchman swatted Bill and knocked him down. When he came up off the floor, Bill grabbed a butcher knife off the table and took a backhand swipe at his opponent. Opened him up like a hog, just under the ribs. For a wonder he didn't touch a gut, but they all spilled out on the floor. A man who was there, telling about it later said, "Godamighty, I didn't know a human had so damn much tallow in his guts as that Dutchman did!"

Tallow or not, the latter must have been a hard case himself. They stuffed his innards back in, wrapped a couple of dish towels around his middle—which makes me squirm, knowing the caliber of dish towels in the average horse camp—laid him in the back of a Model T and took him forty miles or more over country roads to Circle. The doc there sewed him together and be damned if he didn't heal up good as new. I doubt that he got into any more arguments with Bill, though.

The latter had one remark to make about the fracas, "Damn lucky that knife was handy when I come up off th' floor or I'd 'a probably killed th' bastard." Bill had the bark on, for sure.

Gramp had a great friend in Big Timber. He was Irish as Paddy's pig, and a member of the town council. He had definite ideas and no qualms about stating them. One of them was his stand that women should not frequent saloons. "Do you mean good women or bad?" he was asked.

"I mean both of them," was his answer. "The good women because it makes them bad, and the bad women because it will do them no good."

One day, as Gramp was walking down the street, he saw the old gentleman busy as a badger in the middle of his lawn. He was digging a big hole and beside it lay a pineapple. He'd worked up quite a sweat and was glad to knock off and say howdy. They passed the time of day, and then Gramp's curiosity exploded. "What you doing, Pat?"

"Ah," was the answer, "I'm plantin' this thing." Pineapples were in damn short supply in these parts back then.

"What is it?" Gramp asked. Not that he didn't know, for he'd spent some time in Hawaii with kinfolks as a youngster.

"Ah, it's a prisint somebody sint me. Some kind of a boolb or somethin'."

There wasn't much fruit on ranches in the old days. "Two Dot" Wilson was cussing things out to Gramp one day: the wind, the tough winter weather and the dry summers. He ended with a flourish, "Yes, an' goddam a country where dried apples are considered a fruit!"

Even when I was a kid, fresh fruit sure wasn't very plentiful. The only oranges I ever saw until I was pretty well grown were those that Santa used to put in the toes of our stockings. And the peels off those we had to save so Granny could make marmalade. Still, it was hard to beat those dried peaches and apricots that used to come in wooden boxes packed in so tight it almost took a knife to pry one loose. Great stuff to slip in a pocket, if you could get away with it. Prunes, too. Everybody ate 'em, including the pack rats. As the old jingle went: "Strawberries may come and strawberries may go, but prunes go on forever." That was about it, but I still like them. Only trouble now is that they are so damn high-priced. Why, one of these pretty, little, dressed-up packages of prunes costs more than a wooden box of them used to.

There was another staple at most ranches, particularly camps: condensed milk. Maybe there were other brands, but I never saw anything but Carnation. Eagle Brand eventually came along, but it was Carnation that was immortalized in a poem. I have no idea who was the genius that conceived it, but I've heard it all my life.

> "Carnation milk, the best in the land,
> The best damn milk ever put in a can.
> No tits to pull, no hay to pitch,
> Just punch a hole in the sonofabitch."

There's another thing I have vivid memories of. It was a common practice when milk cows were giving a lot in the spring to make butter and stockpile it for use later when cream was harder to come by. There wasn't much refrigeration then, but if butter was salted heavily and put in the springhouse or the cellar, it kept pretty well—either that or everybody just got used to the damn stuff. I never did. As it got rancid, sort of steady by jerks, it was horrible! I remember all too well stopping at a ranch, being invited to eat (naturally) and getting my mouth all fixed, for the lady was known for her fine food. She had just baked, and there was chokecherry jelly on the table—which puts me in mind of the Norsk who was asked to pass the jelly. "Yeesus," he said plaintively. "After t'irty jears Ay learn to say jam, an' now dey call it yelly."

Anyhow, when the lady uncorked the butter, I flat lost my appetite. Even the flies pulled out. Some people claimed it was good to cook with when it got rank, but I could taste it even then. I never got over it, for right now I'd rather eat margarine than butter. Ten to one.

Then there were those eggs I had to fish out of a crock full of that damn, stringy waterglass down in the cellar, for we used to put by eggs, too, when the hens were laying well. I won't eat a boiled egg to this day—God knows what I might find when I open it.

Years ago, when we started the dude ranch in '22, our nearest neighbors as the crow flies were old Mr. and Mrs. Fischer up at the head of Devil Creek. A good ten miles by road from us, anyhow four by saddle horse, but that crow had only about a mile-and-a-half trip. He'd have had to fly over some salty country, too. I have no idea how they managed to exist, for they were tucked back against the timber where the snow must have been a booger. Now nothing but a few rotting boards mark where their buildings were, and sagebrush has taken over the ground they farmed—and there wasn't very much of that. Each time I ride by their old place I marvel at the raw courage and unquenchable spirit of those two. By God, they make those individuals of today, who do so much crying for county, state and federal aid and whine about tough times, look pretty sorry. The Fischers belonged to the old breed! They didn't have much but what they did have they were glad to share with whomever came by. And they loved company.

Their son, Paul, helped his dad, for Mr. Fischer couldn't have done everything alone, crippled as he was. Still, he found time to make a home brew that was as clear and good as a beer can be. He and his wife must have picked an awful bunch of chokecherries, too, for he used to make a lot of wine.

I think it was the summer of '24 when the old couple had their golden wedding anniversary. When the big day arrived, we loaded an easy-gaited, plumb-gentle pack horse with all sorts of goodies Mother and the cook had made, and the whole outfit, dudes and all, rode over to help the Fischers celebrate the occasion.

They were tickled to death. As I remember, two daughters had come for the day—though maybe one of them was Paul's wife. I was just a kid and didn't get everybody straight. Right away Mrs. Fischer hurried into the kitchen and started bustling about to get food for these unexpected guests. She eased off when the pack horse was unloaded, but the old gentleman insisted on a trip to the root cellar for something to wash the food down with. He made several trips before the afternoon was over.

Everybody had all they could eat, and it damn sure wasn't just what we'd brought along. I'm afraid, too, that the inroads made on the beer and wine must have made the coming months a little dry for them, but the two kept insisting on just one more. Of course, I was a button and didn't get any. Oh, maybe a taste or two.

Shortly before we all had to leave, my Aunt Allie got out her harmonica. She played "Long, Long Ago," and those of us who knew the words sang. I can still see those fine old people, holding hands as they listened with tears trickling down their cheeks. I never hear that tune or ride by the place they lived that I don't think of it. It was lovely.

When we finally left, the pair stood on the porch and waved until we dropped over the ridge into Big Timber Canyon. We all felt good as we rode home. We'd contributed to the happiness of two nice people, and a little bit of it had rubbed off on each of us. It has a way of doing that.

When Barbara and I wintered up at the dude ranch for a few years when we were first married, the Fischers were long gone. Our nearest neighbors were down the canyon some four miles. They had a little place at the foot of

Lookout Mountain and perhaps a half-section of range on the bench above. I never heard his given name. His wife, Daisy, called him "Dad," but he favored "Buck," so that's what I always remember him by. At that time he was in his late eighties, and I figured she was a little younger. She looked it, anyhow. They were all right.

He'd lost a good ranch to the bank—I've heard that the man that took over the place damn near paid for it the first year with the grain crop old Buck had already seeded. Anyhow, when the two moved over under Lookout, all they owned was the wagon they were in, the team hitched to it and what they had in it: a cookstove, a few utensils, a little furniture and damn few personal belongings. There was a log house of sorts at their new location, an old barn and a weeney-edge shed or two. They got by though I don't see how in hell they did.

As time went on, he got a little patch seeded, built a ditch from Amelong and used to raise a nice crop of first-class horse hay, and he would put it up all by his lonesome. They had a fine big garden, she canned all sorts of things, including jams and jellies from wild fruit and a berry patch they'd planted. They kept a milk cow, too, and a handful of range stock.

They were quite a pair. She was way in the lead of most people when tongues were handed out, but she'd only beaten him by a nose. I don't remember her ever leaving the place, but every now and then he'd saddle one of his team and ride up to pay us a visit. I might add that his teams were light but pulling devils and sort of goosey. Neither of them was what you'd fit an old man out with as a saddle horse. However, I never saw any snow on his back when he came visiting.

He was tough! I don't believe I ever saw him anywhere but outside when I rode up to his place. Always doing something. I bought a load of hay from him and came down with Kelly and Jack with a rack on a bob. He loaded the hay while I pitched it on. It was a bitter February day, but when I suggested that he ought to be in the house, he told me, "I need exercise, an' I'm glad to get away from th' damn racket in th' house, anyhow."

One time I mentioned that he should hire someone to help him. "Hell," was his answer, "I did once, but it was more damn trouble follerin' him around to see he did things right, than to do 'em myself."

Somehow he came up with an old Chevy "coop." Since he couldn't drive, it usually sat around where he could admire it. When the weather looked good, he'd sometimes ride up home to ask if I could drive him to Big Timber the next day to get some things he needed badly. I'd find out what Barbara wanted—damn little, since we were about broke most of the time. The next morning I'd ride down, he'd fill the manger to overflowing for my saddle pony, and if the road didn't look too bad, Buck and I'd go to town.

He never seemed to bring back much from Big Timber. A sack of flour maybe or some cornmeal. He raised nearly everything they used. I think he went in mainly to sit proudly, ramrod straight, whenever we passed anybody or drove down the street. Then, when we parked, he buttonholed bystanders and regaled them with stories about *his* "coop." There was another thing he brought back. In fact, I'd venture to say that it probably was what he had "needed," and he took it home with him both inside and out. Anyhow, when I hunted him up to head home, he had invariably drink taken and was in a jovial mood. He'd bid a spirited farewell to everybody in whatever bar I found him, buy a bottle (usually a pint) and then sleep soundly all the way to his place.

If we got back early enough, he'd insist that I give him a driving lesson. He'd get behind the wheel, I'd put her in low, and away we'd go around the hay patch in front of the house. There were a number of glacial boulders in the field, some of them a good deal bigger than the "coop," and he'd concentrate grimly on missing them, though I was beside him and poised to avert a collision if need be. Daisy would come out and entreat him to stop before he killed himself. He paid no attention, even when she switched to louder and more forceful orders. Buck happened to be on her side and not fifteen feet away as we made a circle by the house one day when she called, "Dad, you stop that thing and get out. Right away, you hear me?"

The old man poked his head out his window and bellowed, "Git in th' house, woman. This here's man business!"

His wife burst into tears, threw her apron over her head and fled. "You got to tell 'em," he growled.

It made me a little uneasy. I felt sort of as though I was breaking up a home. Incidentally, I don't think Daisy ever figured out that he had been in the sugar. I imagine the old devil left his bottle in the car until he could smuggle it out when he went to milk or something. That wouldn't have been hard, for she wouldn't go anywhere near the "coop." Scared to death of it.

His wife was always after the old fellow to sell out. She had a rest home all picked somewhere near the Blue Mountains or the Grande Ronde Valley out in Oregon, if I remember right. If they sold, they'd have the money to pay for it, and for their funerals, to boot. I don't know why she favored that region, maybe she'd come from there originally. Buck was dead set against it. "Hell," he told me, "I don't want to go some place where I got to set around on m' ass with a bunch of ol' duffers all th' time. I got to keep busy."

A few years later, after we moved to Melville, their house burned down. They lost everything. The neighbors got together and built them a nice little house, and they tried to start over again. I guess it was just too much for the two, Buck especially, because before long they sold and moved out to Daisy's rest home. About a year later Buck was dead. My bet is that, the tough old man that he was, he

couldn't stand the "settin' around." Especially away from
the country where he'd lived so long.

Yet perhaps it was only fair that Daisy, after years up
near the head of the crick, could spend the rest of her life
enjoying the companionship of her peers. I don't know.
They were a nice pair, and I'm glad I knew them.

When we lived at Melville, we had a neighbor up the
creek that was cut out of the same leather or maybe I
should say rawhide: an irascible old Scotsman, Bill Allen.
He was a good man to have over the fence, but he annoyed
the hell out of me sometimes. The ditch from the Sweet
Grass to our place headed a little above Allen's and ran
along the bank above his house between it and the barn. He
was leery of it washing him out, so every time it clouded up
in the mountains, ten miles or more away, he'd beat it up
and shut the headgate down. If it was really raining, I
wouldn't have blamed him a bit.

One day, when I had my water all set, the ditch went dry.
I got a saddle horse, rode up and found the gate shut
tight—there'd been a little shower that morning. On my
way back, in my upper field I met Bill coming home from
Melville and asked if he'd been messing with my ditch. "I
did thot," he rasped, glowering at me. "I dinna want yer
domned ditch to wash me oot. Ye saw th' rain."

Well, I was tired from shoveling all day, and I blew up. I
stepped down, got him by the shirt front and told him,
"You old Scots bastard. If you were thirty years younger,
I'd take your peanut of a heart, if you have one, out with
my bare hands." I shook him a little, "But I'll forget and do
it anyway if you ever touch that headgate again." Then I
rode off, feeling ashamed of myself.

About a year later, Mandius Thompson, who was on the
ditch with me, and I were cleaning it. We had the job all
done except for where Bill had a little foot and sheep bridge
between his house and barn. I went down to see if we could
move it, and Mrs. Allen said, "Beel isna hame." Mandius

was loath to touch the thing, so we fiddled around shoveling for a while. We weren't getting anywhere when Bill showed up from the brush—where he must have been watching us. I was a little put out when I accosted him, but before I could say a word, he poked out his wire stubbled chin, "Dinna touch me noo. I'm on me ain lond!"

All of a sudden it struck me as funny. I laughed and answered, "I wouldn't touch you on a bet, you damn grouch. But can we move your bridge to clean the ditch?"

There was a pregnant silence. Then I saw a twinkle grow deep in his eyes, "Ah, move the dommed brudge, but mind ye, pit it back as well."

So we did, though Mandius whispered when Bill had gone, "Ay t'ought ve vere goin' to be run out ven you said he vas a grouch."

There were two things about Bill that were especially noticeable. He was always squeaky clean as though his wife washed clothes several times a week. Probably did, for even his bib overalls were spotless. His other feature, and it stuck out like a preacher in hell, was that he figured everybody was out to cheat him somehow. He exuded distrust. Almost every year he'd buy hay from me for his little bunch of sheep, because I always had a small stack in my upper field, handy for him to get to. We'd measure the height, width and overthrow to get the tonnage. He wouldn't figure it out at the stack. Hell, no. He had to go home to do his arithmetic. A day or so later he'd come down to see if my conclusion jibed. If it did, he'd make out a check.

One year we measured the stack as usual, but when he arrived at the house, I had come up with more hay than he. I never worry about figures. My wife does them, so I know they're right. Bill wouldn't show his measurements and backed out of the house keeping a wary eye on us. The next day he was back. A shade triumphantly. He'd had Bordie Green up the creek figure the tonnage, and it was the same as his. "An' everybody knows Bordie's honest," he added, looking askance at me. I agreed as to that, but said they both, by God, were wrong. He appeared again a few days later, oozing confidence. He had been to Ronald McDonnell in Big Timber. I don't know whether it was the name or the fact that Ronald was a banker that had caused Bill to seek him out. Anyhow, he and Bill had come up with the same answer—and it was still not right. "But McDonnell canna be wrong," I was assured.

"All right, Bill. Let's go measure the damned stack again. One of us made a mistake," I finally said.

"Well, wasna me, but we'll do thot."

We went up to the stack, spiked the end of the tape to the ground so we could read the footage together and got the measurements. He took his home as usual. Next morning there was a knock on our door just after daylight. It was Bill. He wouldn't come in. Just hunted through himself and came up with a check which he handed over without a word. It was for the amount I had figured. I looked up from it, and he was already leaving. "What happened, Bill?" I called.

"Th' wudth an' th' overthrow. I muxed them up th' first time," and he left.

I felt sorry for the old fellow. He was so damn embarrassed, and I don't think he *knew* how to apologize. He bought hay in later years, but the incident was never mentioned by either of us. I didn't want to hurt his feelings.

Barbara and I thought a lot of Bill. He was cranky, but underneath the prickliness there was a damn good man. When he'd let him out.

19

RODEO

This is the rough draft Spike wrote on the three days preceding his death.

THERE ARE SO many things that I recollect from nearly forty-five years of rodeo that it's hard to keep them in place. On top of that are the stories I've heard about different ones. I always liked the one in White Sulphur Springs. Its date was always about two weeks after our Melville goings-on, and maybe it was because I had shot my wad, as far as responsibilities were concerned, putting on the Melville Rodeo, so I went to the Springs without a care in the world. To contest and to celebrate. That show was pretty much in the lead as far as general all-around excitement goes, in my book. I never was any hell of a hand, but I doubt that anybody had more fun than I did. That's what I remember, the fun. Hell, what a man won was pure gravy. Maybe the reason was White Sulphur, like Melville, had a special joie de vivre. It was a cowboy show. I especially enjoyed it because I wasn't running it. Secondly, since it was along towards the full equinox the weather was plumb unpredictable—the only thing you could bet on was that it would be bastardly. Windy as all hell, or hot, or rainy, or cold, and all too often there'd be a snowstorm. Thirdly, the town was always booming and wide open. Faro to Craps, and everything in between. You'd be likely to see some old booger you'd taken for a sheepherder cash in a thousand or two in chips, walk out and climb into a big Packard and take off for his ranch, for the ranches around town were big ones. Good old-time ranches and the men that owned them were sure enough well-to-do.

First time I was there they had a goat roping, which was frustrating. It was windy, there was a big weighted flag in front of the stands, and if it bellied out over three inches your goat would seek sanctuary behind it and set the brakes. So the goat roping was changed to calf roping the next year. I began to realize what a fine show it was that time. I'd come up alone, pulling my rope horse, Villano— which means anything from villain to sonofabitch in Mexican. Incidentally, I talk some Spanish, and it tickles the hell out of me to see these big old "macho" boys. Hell, "macho" means jackass in Mex, a breeding jack mule.

I had a small flatbed truck with some hay on it, and a sack of oats. When I got in I entered, then went up and camped as everyone had to, at the grounds. I ate downtown, circulated around a little greeting friends, but she got pretty damn gamey so I went on up to bed. Hell, there'd been a bickering or two where fistfights gave way to knives before the law arrived. I rolled my bed under the truck, got in and was dozing off—there's nothing makes a man sleepier than to hear his saddle horse munching contentedly right handy, and Villano was tied to the far side of the truck.

About then a car drove in, circled so its lights hit all the parked riggings. When they hit mine, it stopped, the lights went out and I heard the door open. By that time I was wide awake.

I always made a habit of carrying a gun when I rodeoed. A .44 single-action Colt that a retired Border Patrolman gave me nearly forty-five years ago down in Arizona. A gun squares up a lot of unevenness at times though Gramp told me, "Never pull a gun unless you damn sure aim to use it."

Well, I was fixing to do just that, after the excitement I'd seen downtown. I slid out of bed, Villano snorted softly and a man materialized out of the darkness on my side. I gave him a second or two to say something, then I poked my old gun in his belly, hard, and rasped, "Get th' hell outa here!" It was quiet, and you could have heard that six-shooter click for half a mile. He got!

I shoved some more hay where my horse could get it and turned in again. In about ten minutes all hell broke loose. Two or three cars with blinkers going boiled through the gate, pulled up to my camp, and a string of men got out. I sat up to see what was going on. Seemed like everybody had a flashlight, and all of them were focused in my face. I couldn't see a damn thing, but I could hear an excited voice claiming, "That's him! That's him!"

Finally the beams shifted, and I could make out a highway patrolman standing by me with a gun in his hand. "Did you pull a gun on this man?" he asked, and his tone was not very friendly.

"Damn if I know," I came back. "I can't see him, and it was dark anyhow, but I damn sure put the run on somebody with my six-shooter a while ago."

"Was it loaded?" was the next question.

"Hell, yes, I might as well have a club as an empty gun."

"Don't you know concealed guns are against the law in Montana?"

"Look there," and I pointed to where that big old single action lay beside the head of my bed. "Godamighty, I couldn't conceal that in anything smaller than a gunny sack."

Things hung fire a minute while the law studied, and then another car pulled up, a flashlight beam showed the occupant to be a patrolman from the Melville country. "Hell," was his comment, "that's Spike Van Cleve. I had dinner at his ranch a couple of weeks ago after the Melville Rodeo."

The law was having a low voiced conference when the old boy in the background who'd been repeating, "That's him, all right," asked, "When you goin' to arrest him, damn him?"

"Ah, shut up," said the man from down home. "You're just damn lucky you didn't get shot. The horses these cowboys are carrying are damn high priced, and the boys are a little touchy about anybody messing around with them. Beat it!"

"I'll take him with me," the other patrolman told us and left.

We talked a little after he left. Things quieted down, my horse went back to his hay, and I rolled in and slept until sunup.

That morning, who should come in while I was eating but the first patrolman I'd run into the night before. "Say, Spike," he greeted me with a grin, "You know that fellow you ran off last night?"

I admitted having some knowledge of the affair. "Yup."

"Poor bastard! He had a girl with him, an' your flatbed with all the hay looked like a fine place to entertain her."

"Hell," I said, "if I'd known that I wouldn't have bothered him. At first anyhow, though I'd sure 'a been tempted awful bad, about the time things were going strong, to stand up, cock my gun and ask 'What in hell's goin' on?.' Then he'd 'a had a pretty legitimate kick coming, but it sure would have been fun to see the two of 'em scatter!"

Anyhow we had a cup of coffee together and a good laugh.

I remember another White Sulphur show. Bob Langston and I were traveling together. Don Sturgeon caught a ride with us, planning to hook up at the show with Jay Parsons and Jack Avery, who were working for us at the time—before the war. Don had been in the sugar, but we laid him out on the back seat and he slept most of the way to the Springs. Oh, he'd come to once in a while, sit up with "I'll match any damn man a fifty," but about that time whoever wasn't driving would poke a bottle into his mouth, he'd down a swallow or two, and go back to sleep. We delivered him to Jay and Jack, but I don't believe I ever saw him again until we'd all gotten back to the ranch three days later.

Avery tickled me. The evening after the first show he told us, "By God! Gotta call m' wife. Promised her I would, an' I better do it now while I'm in good shape."

I remember it real well because the bar had the first dial phone I'd seen in Montana, and it took some figuring by all of us to make it work. Finally Jack got hold of her and told her how much he was missing her. He'd been in the 'dogging, had won a third place day money, and she evidently asked how he'd done. "Jes' fine, Punkin," he told her exuberantly. "Took a third an' am settin' high in th' average."

Evidently the fact that there was no way a third money the first day of a two go-round show could be "settin' high in the average" escaped her, for they talked a while. When he hung up he asked, "How'd that sound to you fellers?" We assured him a preacher couldn't have sounded soberer. "That's good," he told us and wandered off into the crowd.

That morning before the show a guy who had celebrated complained about seeing not two, but three calves when we were running them. The consensus of opinion was that he'd better rope at the middle calf. Maybe the advice was good because be damned if he didn't take part of the day money.

It was several shows later that I damn near got whipped. A man from the Wilsall country was at the Springs. He'd been a damn good hand and rode good horses which he raised. But booze had gotten the best of him. Whiskey Dick had gotten puffy and had such a paunch you'd swear he was riding a muley saddle. Anyhow he came out as a header in team tying, made a good catch, stretched the steer and spun his horse back to see if he could help his heeler any. They made good time, but it was sure a funny sight. I told him as they rode back, "By God, Dick, looked like you were roping with your umbilical cord."

He had no idea what I was talking about and was about to hang one on me. Thought I'd insulted him until somebody explained, and he cooled down. It sure looked like he was, though—the rope coming out of that big belly, with the saddle horn completely out of sight.

You damn sure learned not to mess with the law at White Sulphur. Mike Bengan was the sheriff, and he was so hard that if you hit him with a clunk he'd ring like a bell. His deputies were about the same caliber, and the night-sticks they carried were spokes off the hind wheel of a freight wagon. They all were square guys, though, and pretty tolerant. Up to a point. So long as a man didn't get ornery he didn't get bothered. When the show was over— midnight of Labor Day—it was damn sure over.

A few of us were standing by the front of the saloon, just minutes after midnight, when down the street came Stan Lyman. He was a good kid, a dude at the Dot S Dot, but at the moment he was anything but happy. Over his shoulder was a striped barber pole and a step behind him were a couple of cops, who were taking turns swatting him gently across the britches with their spokes. When he saw us he entreated, "Help me, guys, help me!"

One of the law slapped his seat a little harder. "I told you, damn you, th' show is over. Plumb over, but by God, you're goin' to put that pole back. Git to it."

They looked at us as though we might have some objections, but none of us were eager to tangle with those spokes. We did follow along until they arrived at a barber shop where Stan was made to put the pole back and put in the screws with his fingers. Then he got out his knife and tightened them. "That's fine," he was told. "Thanks for puttin' 'er back. But remember what I said. It'll save you trouble."

As they turned the corner up the street, Hughie Stewart turned to Lyman, "Hurry up, Stan, an' you can get away with th' pole before they come around again—if you want it so bad."

The answer he got, though short, was concise. "You go to hell!"

The law got the worst of it once anyhow, as I remember. There was a hotel in town that for some reason or other stuck out in the street eight feet or so. A saddle bronc rider—I've forgotten his name—had a falling out with the cops, he got loose from them and sought haven in the hotel. They were hard on his heels so he retreated to the second story. At the head of the stairs was a fire hose on the wall. He had just time to take it down and turn her on. When the minions of justice poured up the stairs, he washed them into a heap at the bottom. They regrouped, tried again, same result. The excitement had gathered a substantial crowd, all of whom were cheering the fugitive. Twice more they rushed him only to suffer wet, ignominious defeat. Meanwhile the guy with the hose was having a hell of a fine time. Even got to practicing his aim on light fixtures and things between rushes. Then some smart guy went down in the cellar and turned off the water. That did the business, and they bore him triumphantly off to jail. The next morning we guys took up a collection and paid his fine. The fun we'd had watching was well worth it!

Jim Bostad was a saddle bronc rider, a good one. He also was a jockey. He had a bent towards fisticuffs as well, but being more or less of a runt, usually overmatched himself. One night at the Springs, Jim kept showing up, each time a little more worked over. Curiosity got the best of one of the boys, and it seemed Jim had been having trouble with a bartender. We knew the man, a big surly bastard, and insolent to boot. I don't mind a bartender who knows it all, like most of them seem to think they do, but damn one that is insolent. So myself, Danny Ledbetter, Jay Overman and a couple of others decided to go down and see that Jimmy was given the service the customer deserves at a bar.

Jim walked in first. He was spotted, and the barkeep headed around the end. The rest of us stepped in, flanked

Jim, and the man went back where he belonged. "Give me a bourbon and ditch," Jim told him.

He got it. Downed it as the man put glasses in front of each of us and filled them. "Whoa," said Jim "I'da bought a drink for my friends, but your booze is too sorry. Offhand, I'd say your horse had distemper," and he walked out. We followed.

Jim had made his point, but from then on he was careful to take his business somewheres else.

I was team tying director for the MRA and was at the office threshing out some of the problems when Bud Story poked his head in the door. "Got your checkbook with you, Spike?" he inquired brightly.

I looked up from the table. "You bet."

"Well, you're damn sure goin' to need it," he assured me. "They tried to put Buckshot in your shower an' she tore th' pipes loose. Better hurry for it's gettin' awful wet."

I hurried down to the motel and found Buck and three big young contestants, all soaking wet, trying desperately to patch a series of spurting leaks with chewing gum, toilet paper and towels. They weren't getting the job done. I went over and told the lady at the office what had happened. She turned the water off and told me, "I'll get a plumber down right away. When he's gone, we'll get the place mopped up for you."

"Thanks, Ma'am," I told her, "but I want to pay for the plumber."

"Forget it," she grinned. "We figure on things like that happening. It's rodeo time."

Getting a meal was invariably quite a problem, for there were only two cafes in town. They did their best, but sometimes a man would have gotten fed faster if he'd driven down to Martinsdale to eat.

After the show one evening Buckshot, Story, the Okie and I went to eat. It had rained hard all day, and the floor of the cafe had about half an inch of mud, manure, cigarette butts, and what have you spread evenly on it. We finally got a booth and were ordering when a young fellow stepped up and asked if he could sit at the end of the table. He was a good hand, from way-back at the head of the creek, and sort of bashful. We said sure, so he rustled up a chair and gave his order. We were about done when his steak arrived. Fortunately. Man, but the boy perked up! The table was crowded, but he set the platter in front of him, picked up his fork and knife, started to cut a bit off, and the whole works tipped off the table, missed his knees and ended upside down on the floor. With a stricken expression he leaned down and retrieved things. The steak he put, clean side down on a napkin, cleaned the platter with another, put the meat back and scraped off with his knife the gunk from the side that had hit down. It took a little doing and a little time, but he got her done. Though there was no messing with the french fries. Finally he put the napkins he'd used on his knife and platter on top of the french fries between his feet, cut a chunk of steak and started to eat. He glanced around the table, sort of like a kicked pup, and looked down again. Buckshot was swallowing pretty hard and even the Okie didn't look very shiny. Don't imagine I did either, but it never fazed Story a bit. "Spike," he remarked judiciously, "now that's what I'd call a hungry cowboy!"

Empty saddle

HOME

Out toward the setting sun there's a land
Where the brawling Yellowstone
Drops snakelike through the foothills
Where Cheyenne and Crow were known;
There's a strong voice calling, calling,
My son, oh come you home.

A far flung, spreading rangeland with
Its grey rims in the sun,
Where the song of the coyote lifts to the stars,
And the slim, fleet pronghorn run;
There is where I wish to be,
For I love them, every one.

Where the winter blizzard whips the snow
On the slow herd drifting past,
And the saddle pony trudges through
The whirling, icy blast;
It's all bred into my heart and soul,
God grant it will always last.

There the sound of the cowboy's voice
As he sings to the bedded steers
And the friendly sounds from a roundup camp
Come faintly to the ears;
They all belong to a son of the range,
A heritage great and dear.

Where the smoke from red hot iron on hide
Mounts in billowing swell,
While the struggling cattle bawl in fear
And the dusty 'punchers yell;
The brand of it all is on my soul,
Easy to read and well.

When my last circle is ridden and I
Have hung up saddle and rope,
Put me out on a short grass ridge
With a sunny, southern slope,
And the stone that rests o'er my sleeping head
Will carry this verse, I hope.

"The riders are home from the milling herd
Which fades in the darkening day;
The ponies drink at the cool, clear creek
And slowly graze away;
While far from the sage blue circling hills
Comes a coyote's evening lay;
All is rest and peace, at last,
For a son is home, to stay."

—Spike Van Cleve